IMPOSSIBLE LANGUAGES

IMPOSSIBLE LANGUAGES

Andrea Moro

THE MIT PRESS
CAMBRIDGE, MASSACHUSETTS
LONDON, ENGLAND

Published by arrangement with The Italian Literary Agency.

This book was set in Futura Bold and Stone Serif by Toppan Best-set Premedia Limited. Printed and bound in the United States of America.

Library of Congress Cataloging-in-Publication Data

Names: Moro, Andrea, author.
Title: Impossible languages / Andrea Moro.
Description: Cambridge, MA : The MIT Press, [2016] | Includes
    bibliographical references and index.
Identifiers: LCCN 2016001184 | ISBN 9780262034890 (hardcover : alk.
    paper)
Subjects: LCSH: Grammar, Comparative and general--Syntax. |
    Biolinguistics. | Linguistic analysis (Linguistics)
Classification: LCC P132 .M66 2016 | DDC 401--dc23 LC record
    available at http://lccn.loc.gov/2016001184

10 9 8 7 6 5 4 3 2 1

To FB

CONTENTS

ACKNOWLEDGMENTS

Never agree to write a short book on anything: the task turns out to be more difficult than most of us expect, especially because it challenges us to expose our deepest thoughts. But this is also arguably the most useful type of book for readers because, more than any other type, it may reveal the author's new clothes. I have many people to blame for this: Noam Chomsky, Stefano Cappa, Gennaro Chierchia, Luigi Rizzi, Ian Roberts, Alessandra Tomaselli, and Richard Kayne. But if it weren't for them I wouldn't be able to share my passion, and hope you'll enjoy doing the same with your friends. A very special thanks to Marc Lowenthal for fostering these ideas with great generosity and criticism. All errors are mines, including this.

The insects inhabiting the surface of our planet today are not substantially different from their remote ancestors who flourished six hundred million years ago. Since the first specimen appeared, their speck of a brain has shown itself so well suited to the problems of the environment and predators that it has not had to submit to the capricious game of mutations, but owes its evolutionary immobility to the perfection of the primordial model.

—Rita Levi-Montalcini, *In Praise of Imperfection*, 1987

1

LEARNING FROM THE IMPOSSIBLE

To define the class of possible human languages: this is the ultimate aim of linguistics. Prima facie, it may seem too limited an objective, but that is not the case. First, a major and preliminary question arises: Do impossible human languages exist at all? What is a language made of that there can be an impossible one? Here we have already reached the core of the problem. There are at least two ways to consider a human language. One is to describe it as a set of discrete primitive elements (to put it broadly, a dictionary of words, each constituted by an arbitrary association of sounds with meanings) and a set of rules of combination (its "syntax," to use a word coming from Ancient Greek meaning "composition") that use these primitive elements to generate a potentially infinite array of structures, which are then interpreted at the relevant interface (minimally, at the articulatory and conceptual interface) and thereby convey and communicate complex meanings. The other way to describe it is from a purely physical point of view. From this perspective, a human language lives in two different environments: outside our brain and within it. When it lives outside our brain it takes the form of mechanical waves of compressed and rarefied air (i.e., sound); when it lives within our brain it takes the form of electric waves, which is the code neurons exploit to exchange information when they perform cognitive operations (i.e., brain activity). So when we ask whether an impossible language exists we are in fact asking a twofold question: a formal one (concerning rules) and a physical one (concerning matter). The aim of this book is to focus on both sides of this coin and ultimately argue for their possible unification. In doing so I will consider several different aspects of human language, always keeping in mind the fundamental goal of linguistics I have highlighted, and rely for the empirical considerations on a few experiments I designed with different teams and others that have been recently published (see Moro 2013, 2015, and the references cited there).

It goes without saying that, besides having these two sets of properties (formal and physical), human languages must be endowed with some other general properties that make them not only possible but usable. These properties are trivial, though. For example, a human language must be rich enough to carry all the meanings we can think of; it must be capable of allowing logical inferences and coherent pragmatic interpretation; it must lend itself to communication; and, ultimately, children must be able to spontaneously acquire it within the first years of their life. But these properties still follow from our major question—whether impossible languages exist. If we were biologists, we would not hesitate to claim that there are impossible animals: an animal that produces more energy than it absorbs, for example, or an animal capable of indefinite growth. We could make such a claim because all organisms are constrained by physical laws, like entropy or gravity, and it would be relatively easy to formulate the notion of "impossible organisms" to yield a compact description of the animal kingdom (the literature is vast, ranging from Thompson [1917] 1961 to Edelman 1988 and Wesson 1991, among others). With languages, however, the situation is much more complicated, and all the more so from a physical point of view, for it is hard to think of a physical law or an equivalent of a physical law that would make a language impossible. If anything, this rather suggests that the notion of "law" would not only be of no use, but would not even be empirically plausible when it comes to language. There are, of course, properties that we automatically dismiss as being able to influence a human language: for example, we would never think that a language might be sensitive to the temperature or the speed at which it is spoken; this would amount to establishing an unjustified link between a language's formal constraints and its physical ones and it would be as crazy as linking the type of paper used to publish a novel to the personality of the characters involved.

Let us reformulate what we claimed to be the ultimate goal of linguistics and say that it coincides with answering the question of whether impossible languages can exist, both in the formal and the physical sense. As far as the formal sense goes, the hypothesis that there may be impossible languages has not always been considered plausible within the Western tradition. With the exception of some speculations by medieval philosophers like Roger Bacon, a very different assumption has prevailed for centuries. For this reason, the Tower of Babel has not just been considered to be a vivid biblical image, but also an allegory for what many have regarded as a defining property of human languages: a source of total confusion, with Babel the prototype of all possible confusion. From this perspective, there are no impossible languages, simply because there are no formal principles to which languages must adhere; they can follow any conceivable rule, freely and without bounds. Interestingly, this has not just been a popular interpretation, but one that has also reflected the dominant scientific viewpoint as late as the 1950s—one explicitly expressed by Martin Joos (1957, vol. 1, 96), for example, who stated: "Languages [can] differ from each other without limit and in unpredictable ways." Notice that this modern vision of Babel was not accidental. It followed from the idea that formal cohesion in any set of words endowed with a combinatorial rule system is a sufficient requirement for it to become a human language as long as it reflects certain statistical patterns, without any further requirements needing to be met (Shannon 1948): any human language of the past, the present, and the future is just the linear adjacency of statistical regularities usable for communication. This belief about language structure and communication was in turn ancillary to another possible, even more fundamental, point of view, according to which the birth of the human mind—in the individual and in the species, which is to say, ontogenetically and phylogenetically—was reduced to a stimulus-response reaction.

Sometimes, a look at the place and the time in which ideas are born and defended is useful. The 1950s—the time when this view of language was defended—were the years Eric Hobsbawm has described as the "golden era" of science, years in which the immense, hectic, and collective effort in scientific research during World War II was first freed from the need to construct weapons or defense systems and could focus on issues such as the nature of the basic molecules of life, the creation of alternative sources of energy from atoms, or the treatment and transmission of information. As for the place, the United States in the 1950s was the country in which this research could develop better than anywhere else, with the work being done in Cambridge, Massachusetts, of particular importance. The account of Yehoshua Bar-Hillel (1965, 294) describing the atmosphere at MIT's Research

Laboratory of Electronics, one of the major research centers at that time, is illuminating and dramatic:

> There was a ubiquitous and overwhelming feeling around the Laboratory that with the new insights of cybernetics and the newly developed techniques of information theory the final breakthrough towards a full understanding of the complexities of communication "in the animal and in the machine" had been achieved. Linguists and psychologists, philosophers and sociologists alike hailed the entrance of the electrical engineer and the probability mathematician into the communication field.

Humans, in fact, disappeared from the scenario: animals and machines were the only protagonists in the arena, contradicting for the first time in centuries the Cartesian assumption that human language was unique with respect to those of all other animals. As always happens in science, when someone claims to be close to a full understanding of a natural domain—to say nothing of a theory of everything, whatever that may mean—something else unpredicted and unexpected happens, necessitating a new start to the process of understanding, albeit at a higher level of comprehension. A similar thing happened to the science of language and is still happening today. Interestingly, what happened coincided with the exact opposite of the dominant view, namely the birth of the idea that human languages are by no means unbounded, but instead share a unique, intricate set of general principles with relatively little variation, producing a surprising constellation of language properties now recognized by all schools of linguistics, and by typologists and generative grammarians in particular (see Graffi 2000 for a history of the school of linguistics in the last two centuries). All this was developed at the level of formal properties of language, since at that time there was no real hope of getting direct access to the other level of possibility, the physical one, although there were already some hints that one could not account for the first level without having an idea of what happens at the other.

One of the fundamental steps toward a biological understanding of the structure of language and the characterization of the class of possible human languages was taken, not unexpectedly, by a psychologist working in the same intellectual environment as the environment that prevailed at MIT, Eric Lenneberg. In his foundational treatise on the biological foundations of language, Lenneberg (1967, 2) obviously felt obliged to highlight the difficulty he was facing:

> A biological investigation into language must seem paradoxical as it is so widely assumed that languages consist of arbitrary, cultural conventions. Wittgenstein and his followers speak of the word game, thus likening languages to the arbitrary set of rules encountered in parlor games and sports. It is acceptable usage to speak of the psychology of bridge or poker, but a treatise on the biological foundation of contract bridge would not seem to be an interesting topic. The rules of natural languages do bear some superficial resemblance to the rules of a game, but I hope to make it obvious in the following chapters that there are major and fundamental differences between rules of languages and rules of games. The former are biologically determined; the latter are arbitrary.

Interestingly, at least from a historical point of view, the position against which Lenneberg took a stand is not too far removed from that of François Rabelais (1973, vol. 3, 19), the French Renaissance intellectual, who wrote: "It is misleading to claim that there is one natural language; languages arise from arbitrary impositions and conventions amongst peoples." From Lenneberg's point of view, however, there could clearly be impossible languages because he had collected empirical evidence that languages do not consist of arbitrary and cultural conventions. He based his conviction on clinical data: he observed that in cases of language loss (aphasia) due to focal lesions (trauma, tumors, or vascular accidents), the reacquisition of language was more rapid, more complete, and followed the same path as first-language acquisition in individuals who experienced the loss before puberty. Since puberty cannot be considered either an arbitrary or a cultural convention, the conclusion that the same applied to language was inescapable. It is very important to emphasize that Lenneberg referred to *languages*, not just to *language* in general—that is, to the specific implementations (languages) of the general capacities of humans to communicate verbally (language). What this means is that it is not just the general capacity to communicate—something that, mutatis mutandis, is common to many, if not all, living organisms—that is neither arbitrary nor conventional, but the specific format of each and every code manifested by each and every human language.

Something else that influences the development of our understanding of human language, besides any philosophical or ideological bias, is the lack of animal models and, by extension, the lack of possible experiments on animals. One important point must be emphasized here: virtually all animals use language and can communicate. In fact, if we were to reduce our use of the word *communication* to the simple exchange of information, we can safely admit that even plants communicate (Baluška, Mancuso, and Volkmann 2013). Moreover, animals are clearly capable of symbolic communication: the urine spraying used in territorial marking by felines is a clear case where an object (the smell of urine) is interpreted by other cats as standing for something else (the control of a certain territory). What is crucial, instead, is that humans are the only animals that can recombine discrete elements (words, which obviously include symbolic elements) in such a way as to provide new meaning, depending on how the combination is made. Any English speaker asked to combine three words such as *Abel*, *Cain*, and *killed* can combine them to produce *Cain killed Abel* and *Abel killed Cain*. These are two sentences with opposite meanings—in other words, truth-value conditions (the conditions of the world that make a declarative statement true or false). Same words, different positions, different meanings: syntax is the fingerprint of every human language, and only human languages bear this fingerprint. Even chimpanzees, our genetically closest relatives, have been shown to be incapable of exploiting syntax to produce a novel meaning in a famous experiment involving a chimp learning sign language to avoid difficulties related to the articulatory apparatus (Terrace et al. 1979), and the same has been proven true for gorillas in an ecological context (Genty and Byrne 2010). The advantages we gain from syntax are incalculable and probably reflect the uniqueness of the human mind, especially since it can be argued that some form of syntax is manifested in other cognitive domains like mathematics and music (though they are significantly different in their repertoires of primitive elements and in the types of rules of combinations). Animals only have a repertoire of fixed messages, and an exiguous set of messages at that. To put it simply: humans have dictionaries of words; animals have dictionaries of sentences.

In this book I examine the evidence for the hypothesis that impossible human languages exist, and outline how the very notion of impossible languages has helped shaped research in the field. I should be clear, though: any reflection on language is a delicate matter. Every civilization has expressed its views on language; one could even say that language is something of a "Homeric question" for humanity: a question indirectly reflecting the pillars supporting that culture. This is probably the real reason views on language are always shifting, despite our natural tendency to attribute these shifting views to the emergence of new methods or data. Obviously, there are no privileged data in empirical science; only methods can be so. In many cases the journey to a better, if partial, understanding can take different paths depending on who conducts the research, and it can be directed by intuitive capacities that may not always appear completely rational. In the case of human language, things are even more awkward because some hypotheses still considered valid were actually disproved long ago but still persist in the scientific *vulgata*. The scenario the researcher faces is similar in certain respects to a starry night. An untrained mind may not know that some of the light in the sky is coming from dead stars and may make wrong deductions about the relationships between living stars and dead ones. In any case, everything simply depends on what we want to obtain from data. The method itself may not be something that can be evaluated with an absolute and objective metric; a method is more valid than another if and only if it allows us to make more predictions or the same ones with less effort and assumptions—that is, it is simpler—or it allows us to acquire more precision than any other competing method. As for elegance, we can ignore it. I follow Einstein here, who, in turn, said he followed "the rule of the great theoretician L. Boltzmann, that is, the question of elegance should be left to tailors and shoemakers" (quoted in Infeld 1950). If what is at stake is the shape of the world and the shape of language, it is clear that we are not even in a position to understand what the possible predictions are. There is no guarantee that we can solve every problem we are capable of posing, just as there is no guarantee that we can pose every possible problem assuming that they can exist independently of us: our domain of inquiry can only encompass problems compatible with our perception and our rationality. It is even possible that we can solve problems we do not fully understand from an overall perspective, the same way a mouse that gets out of a maze is probably not able to elaborate a general theory of labyrinths. This is not a matter of how complex a problem's formulation may be: Goldbach's conjecture, for example, according to which any even number greater than 2 is the sum of two prime numbers, is formulated in a very simple way, one that even a first-grade student can understand, yet it has remained unsolved since he posed it centuries ago.

Let us approach the problem of impossible languages by first considering some fundamental aspects of what appears to be possible.

2

CAPTURING THE "STEM MIND"

If we take a rational approach to a human language, then what do we expect from a grammar? There cannot be a unique answer to this: it really depends on what one wants to do with a grammar. We may want to learn a foreign language; we may want to know how to deliver a speech; or we may want to explain why languages change. The history of linguistics has seen many differing positions on the role of grammar, and they are not necessarily mutually exclusive. After all, the root of the term *grammar* is the verb *grapho*, which in Ancient Greek expresses the idea of carving a letter on a wax tablet. This is why the term was originally restricted to a set of norms illustrating how to write—more precisely, how to write correctly. But it is also true that even during the classical period, when this prescriptive approach to language was established and commonly adopted, speculative approaches to language flourished and paved the way for successive studies in the field. Needless to say, the first works on language were not limited to those produced by Western civilization; in fact, in roughly the same century that Plato and Aristotle were developing their linguistic views (the fourth century BC), Pānini, a linguist who lived in ancient India, wrote an illuminated grammar of Vedic Sanskrit that, according to Noam

Chomsky (1965, v), "can be interpreted as a fragment of ... a 'generative grammar,' in essentially the contemporary sense of this term." Here, by "generative grammar" Chomsky means an explicit grammar; that is, a grammar that does not allude to the intuition of the intelligent reader but describes each and every rule in a precise and formal level (see Chomsky 2005). Despite this remarkable convergence, however, the two cultures happened to be separated, geographically and chronologically, and Western linguistics could not immediately benefit from the illuminating insights in Pāṇini's grammar—or, of course, vice versa.

We may thus expect many different things from a grammar and, depending on whether we think there are impossible languages, we may get very different results. One important premise must be addressed, though. Whatever the grammar is, it must be constructed by observing data through the same approach any other scientist would adopt when observing data in his or her domain of interest: by proposing progressively adequate descriptions to reach a satisfactory, albeit partial, result in accordance with a predefined goal. For many years, instead, the implicit assumption was that the scientific study of languages, if possible at all, was different from any other: one could adopt a discovery procedure—essentially an automatic algorithm—and apply it directly to the data, as if the regularities of human languages had a different ontological nature than any other regularity. But any approach to languages (and language in general) must follow the same methodological strategy as in any other empirical science. Not doing so would be tantamount to assuming that it would be enough to look at the sun for a long period to obtain the heliocentric model of the solar system; in a sense, this is indeed enough, but before reaching the truth, one has to exclude every competing hypothesis until the only one left reveals itself as the most accurate one. An automatic procedure would hardly be acceptable; in fact, no one has ever expected phenomena to wear their analyses on their sleeves, with the apparent exception of the domain of human languages. Moreover, it must be highlighted that this allegedly neutral view of linguistics was not completely naive, in the sense that it was ancillary to the idea that there was no general principle ruling language structure and that statistical calculations were sufficient to capture its architecture; and therefore was for children spontaneous language acquisition. This of course does not exclude that some domain-general probabilistic learning mechanisms may operate during language acquisition, but nevertheless they operate in a domain-specific space offered by the genetic endowment that is human specific (see Yang 2004 and references cited there). In the next chapter we will get into the details of this genetic endowment and provide formal and empirical arguments in favor of its existence. But just to remain on the linguists' side, to assume an automatic procedure for discovery of language structure would amount to assuming that if one used cameras to record the image of the sun in the sky, some well-designed program would eventually produce a heliocentric theory on the sole basis of statistical data derived from the photographs. Although theoretically possible, it can hardly be a path for scientists to follow, and it was not. But, to refer back to Bar-Hillel's quote in the first chapter, the idea was that there is nothing particularly human in a human language. The crucial fact here is that this point of view implied that there were no impossible languages because there were no formal principles a language could contradict or not be endowed with.

Suppose we want a grammar to tell us how children acquire their language before knowing any other. This was historically and conceptually the key question that constituted the turning point in modern linguistics. Noam Chomsky (1959, 57) summed up this radical change of perspectives: "The fact that all normal children acquire essentially comparable grammar of great complexity with remarkable rapidity suggests that human beings are somehow specially designed to do this, with data-handling or 'hypothesis-formulating' ability of unknown character and complexity." The reductive binomial "animal-machine" that Bar-Hillel referred to as being the focus of the study of communication was dismantled and humans came back into the arena, but as a less studied and relatively unexpected group of humans: children. Before any theoretical discussion could take place, though, there was a fact that was obvious to everybody—probably too obvious to capture many people's attention—but which became central: despite all the degrees of difference among languages and adults' own subjective experience in judging the complexity of a foreign language they are learning, children do all acquire their language(s) in more or less the same amount of time (with some minor variation depending on the richness of the morphological repertoire; see the survey in Kelly, Wigglesworth, Nordlinger, and Blythe 2012). This simple fact immediately suggested an empirical requirement that could help identify a possible language: a possible language is one that must be spontaneously attainable in the same amount of time by any child, independently of all other conditions (emotional, social, historical, environmental, etc.) and in the absence of specific pathologies.

One immediate conclusion was to be discarded, though: that a specific language is somehow directly encoded in our genome. If that were the case, one could expect a child born of parents speaking a certain language to (more) easily acquire that specific language or to have difficulty acquiring a different one, a fact that can easily be contradicted by common experience. Any child can acquire any language in the same amount of time as any other language: at her or his birth, the child's mind is just open to the acquisition of any language. In short, *linguistically, the child's mind is a "stem mind" that is potentially able to specialize in any human language.* It has not been easy for linguists to realize how important this simple fact is, but the resulting model has changed our way of understanding language and the human mind as few other phenomena have. Nevertheless, this view raises a new problem, coming from two competing facts: the same amount of time is indeed required to learn any language, but there are apparent differences among languages. It also gives rise to two questions: How can we identify the "stem" mind capable of developing into any language? What kind of instructions is it endowed with?

The two competing facts just mentioned lead to only one possibility: that—appearances to the contrary—all human languages do share the same structure. More explicitly: they have essentially the same primitive elements and rules of composition, and therefore have the same complexity, although of course there may be variations, such as the obvious ones derived from the arbitrary association between sounds and meanings (technically referred to as "Saussurean arbitrariness," in homage to Ferdinand de Saussure, founder of the structuralism in the early twentieth century; see Saussure [1922] 1975). It is indeed a simple explanation and one that has irritated those who have misleadingly capitalized on differences in grammar to argue in favor of the alleged superiority of one culture over another, whatever may be implied by such "superiority." I am referring to the never dismissed conviction that cultures are exclusive expressions of languages—their "mirrors," so to speak: the richer the language, the richer the culture. Interestingly, no one has ever provided a metric to compute the overall complexity of a language, let alone its culture (see Newmeyer and Preston 2014).

However, if one were to be a devil's advocate it must be said that although this solution is appealing because of its simplicity, the concrete implementation of a theory able to capture such a comprehensive view of language is extremely difficult because it has to account for the immediate observable data, which prima facie do give the impression that languages are qualitatively different. There is no need to get particularly discouraged, though: this is the prototypical starting point for empirical science, the goal of which, as the physicist Jean-Baptiste Perrin put it, is to "explain what is visible and complicated by means of what is simple and invisible" (quoted in Jacob 1970/1974). The standard example given for this is provided by physical elements: the differences among them are apparently irreducible and essentially qualitative. Mendeleev's intuition and the construction of the periodic table, though, have succeeded in the stunning enterprise of tracing the immediate differences back to quantitative differences. A similar effort characterizes modern linguistics: to isolate both the invariant scaffolding and the point of variation for human languages, with the assumption that languages share essentially the same structure as a result of a genetically determined program.

Considering the history of modern science—where the decomposition of complex entities into the interaction of smaller ones, such as in the periodic table of elements or in comparative genomics—it should hardly come as a surprise that microscopic differences in a highly intricate system may lead to dramatic macroscopic differences and thus contradict the idea of a substantially unique structure. In fact, it also crucially depends on the level of observation. From a certain point of view, the

Manhattan skyline is totally different every night because the lights from the windows of the buildings are never exactly the same; nevertheless, there is a sense in which we can consistently say that it is the same skyline every night because we abstract the real data in order to favor other aspects of reality.

Linguistics is still trying to find such a systematic reduction—the equivalent of a periodic table of languages—and the situation remains very puzzling (but see Baker 2001). One of the reasons—it seems to me—is essentially that linguists are not just searching for possible combinations of primitive elements within a well-established schema and by means of a consolidated formalism; they are also constantly and untiringly redefining the repertoire of primitive elements together with the combinatorial rules, not to speak of the formalism (Graffi 2001, Chomsky 2013, Moro 2015). This is somewhat like trying to repair the broken wing of a plane while flying it, all the more difficult if one is also lacking the instructions to do so. Nevertheless, the following definition is now well established in the field of linguistics: A possible language is what survives when a set of options (technically called "the parameters") is chosen within an invariant system of principles (see Rizzi 2009 and Chomsky 2013). This is to say that language acquisition by children amounts to establishing these variable options on the basis of external stimuli (see Piattelli-Palmarini 1989 who witnessed the emergence of this new framework). Experience, obviously, is fundamental in language acquisition, and not only in the obvious case of the arbitrary association between meaning and sense, but because it triggers the entire process from the start. In fact, *one could also define linguistics as the theory of the limits of the impact of experience on language structure.*

Of course, this does not mean that this definition raises no questions or challenges. A major one, for example, is: Why are there variations at all? That is, why doesn't every human language share the exact same structure? Here the comparison with physics is no longer viable because the variations in the periodic elements took place within a well-defined metric of space and energy, whereas in the domain of human language, it is the metric itself that has yet to be discovered. One possibility is to assume that variation is a consequence of complexity, in the sense that a complex system must be somewhat flexible if it is not to be too fragile, and that the set of options are the price to be paid if a system is to be more robust in the same way that small defects are tolerated in a crystal. This, again, could be tested relatively easily with a physical structure; whether this is also true for a formal system like language, though, still lies beyond our knowledge. Certainly, the impact of the "Babel effect" on our species must have been enormous and not necessarily in a negative way. One major impact is that it arguably played a positive role in limiting urban growth in eras in which technology was inadequate to deal with a megalopolis. A common language by which all individuals understood each other—a language without different accents, without different vocabularies, and without different rules—could have led to a catastrophic assemblage of people unable to control resources or any essential aspects of social life. The "Babel effect" in this sense may turn out to be a gift rather than a curse, and in some ways similar to the effect that infectious diseases can have on animals by preventing food shortages. Naturally, this is something that no one could, or should, ever test.

On the other hand, this theory fits very well with the assumption that human languages are an expression of biological restrictions for the very reason that analogous systems have been discovered in other domains. A very interesting system pertains to immunology. A major question characterizing immunology in the second half of the last century has concerned the mechanism behind how antibodies form in order to neutralize antigens (Paul 2013). One of the major landmarks in the understanding of this process has been the assumption that the immune system is similar to language acquisition: instead of producing an ad hoc response every time an antigen enters the organism, nature has provided humans with abundant repertoire of different types of antibodies. Some of them may never play an actual role if they never encounter a disease they can block, but some are already assembled in our body in order to allow the immune response to operate quickly, as first suggested by the seminal work of the Nobel laureate Niels Jerne (cf. Jerne 1985). The same thing happens with languages, and in fact Jerne took inspiration from Chomsky's theory, as clearly indicated in the title of Jerne's Nobel lecture: "A Generative Grammar for the Immune System." As we are designed to acquire many more languages than those we encounter, and more broadly, more languages than will ever be spoken on our planet, so we are designed to neutralize many more antigens than those we encounter or that humans in general may encounter in the past, present, and future. And in both cases, we cannot defend ourselves from this invasion, be it of antigens or sentences; just as our body cannot help but react to an antigen, so it cannot avoid understanding a sentence it is exposed to, once it has been endowed with the code to decipher it.

Ironically, it turns out that the reactions of a human organism to a sneeze and to a sentence may not be that different. They are both decoded on the basis of an abundant repertoire of preformed codes. Of course, producing a sentence is a completely different matter, and there the comparison ends, for there is no creativity in sneezing.

3

SENTENCES AS SNOWFLAKES

Words come in sequences: this is perhaps the only incontrovertible fact about human language. What is surprising is that it is not the most relevant fact. To understand the reason for this, we need to learn how to see such sequences in a two-dimensional way. Exploring a sentence structure is, in some sense, similar to looking at a tapestry: the apparent linear organization of adjacent elements of the structure is very misleading and this is due to external conditions imposed on communication from the way our body is designed. Seen from the front, a tapestry looks like many colored dots, one next to the other, forming some image. But if we look at the other side of the tapestry, we discover a complex structure of woven threads that makes almost no visual sense. Individual threads emerge on the surface and then dive back down, disappearing behind the tapestry while creating an image on the front. In a way, syntax resembles a tapestry: if we look at it superficially, it appears as a simple row of words, arranged next to each other in a sound and coherent way. But if we manage to look at it '"from behind," we discover the hidden and intricate structure that connects the words at a distance. Understanding the structure of this tapestry will allow us to deepen our understanding of what counts as an impossible language; in fact, this will allow us to capture the essential fingerprint of human language.

Consider a simple case: a sentence made of two words, such as *John smiles*. The two words are connected by virtue of many features. Let's consider two relatively simple ones. Some of the features combine well when they mismatch, some when they match. For example, the fact that the two words belong to two different classes—a noun and a verb, respectively—makes them fit together. If they were two nouns or two verbs (say, *John Mary* or *smiles eats*), then they would not. On the other hand, the fact that their "number"—the fact of being either singular or plural—matches also makes them fit together; if they were different (say, *John smile* or *they smiles*), they would not be properly connected. Syntax requires that all features combine appropriately, in accordance with properties that are common or specific to each and every language, such as the ones just mentioned. When two or more words combine properly, we say that their syntax is correct or harmonic (remember that *syntax* etymologically means nothing more than "combination"); if the combination pertains to morphological features, such as the distinction between singular and plural, or the different inflections of verbs (as in *paint, paints,* and *painted*), we name it *morphosyntax: morphos* in Ancient Greek meant "form" or "shape" and morphology is the branch of linguistics which studies the "form" of a word and its building blocks. Technically, whenever two words combine by virtue of some feature we say that a dependency is established between them. Let's now focus on the sequence *John smile*, which is an incorrect combination—as if the linguistic "glue" produced by number failed to keep the correct words together even if their category, a noun and a verb, are candidates for harmonic combination. In other words, this is a morphosyntactic error. Apparently, there is no remedy other than to inflect the verb (or alternatively to get a plural subject). Surprisingly, changing these elements is not the only way to rescue the wrong sequence: if appropriate new material preceded these two same words, the offending sequence would become fully acceptable, as in [*Mary and*] *John smile*. This fact is perhaps too simple to be of interest to many people, but it is nevertheless the sign of a fundamental structural characteristic of human language code. Before revealing it, let's consider a more complex example. Take the sentence *Mary wants to describe herself*, which is a syntactically well-formed sentence, in contrast to a sentence like *Mary wants to describe themselves*. Again, the ungrammatical sequence can be turned into a perfectly well-formed one, provided that it is preceded by other words, as in [*Who do you think that*] *Mary wants to describe themselves?* As in the former case, this sequence becomes fully acceptable if something is added to a sequence that can then contain the addition. In this case, the "solution" is even more interesting, because it implies that the pronoun which *themselves* refers to—namely, *who*—has been far removed from the canonical position where normally a pronoun would occur in the sequence. These types of pronouns involving -self/-selves (called "reflexives") must find their antecedent locally, that is, in a position which is close enough to their antecedent. Consider, for example, the following contrasting sequences: *They think that Mary describes themselves*, which sounds ungrammatical, versus *They think that Mary describes herself*, which sounds perfect. Unless one wants to give up the intuitive and powerful assumption coming from the latter contrast that the pronoun can only have the closest subject as an antecedent, in a sentence like *Who do you think that Mary wants to describe themselves?* one must assume that *who* is pronounced in a different position, one where it plays the role of the subject of *describe*, as is clear in an affirmative counterpart of the same sentence—for example, *You think that Mary wants the pupils to describe themselves*. Putting all these scattered data together, to understand why the original sentence, namely *Who do you think that Mary wants to describe themselves*, is grammatical, one can represent it in a way such that the antecedent *who* is closer to the pronoun; as a first approximation, one can render it as follows: *Who do you think that Mary wants ~~who~~ to describe themselves?* where the line striking through the pronoun *who* simply means it is not pronounced there. Needless to say, the reason for this phenomenon has been debated since this discontinuity was discovered in the early twenty-first century (see Graffi 2001 for a historical point of view and Chomsky 2013 and Rizzi 2015 for the most recent and comprehensive perspective). Whatever the explanation for this phenomenon, it is clearly a characteristic— sometimes referred to technically as *nonmonotonicity*—exclusive to human language. It is not shared by other semiotic systems, by other codes like the genetic code (see Berwick 1996), or by mathematics, and it is the one compelling reason for concluding that a pure linear sequence of words is not sufficient for capturing the rules of human languages. The very fact that a word far from other words in the same sequence can affect them is in itself enough for us to be able to conclude that we need at least an extra dimension to represent syntactic relations properly. The discovery of this peculiar tapestry structure has been at the very heart of research in linguistics since the early twentieth century and has led to some major advances in the understanding of the human language faculty. We will look at two more properties characterizing syntax, both of which will make the notion of an impossible language more concrete and clear. Note that the preceding example demonstrates that there is a phenomenon that cannot be explained in terms of a linear sequence; it also implies a much more powerful generalization: Given that the linear order is a one-dimensional space, *no syntactic phenomenon can be explained without referring to a two-dimensional space.* This will become clearer by the end of this chapter. Why a two-dimensional space is enough is a rather murky fact, probably related to the binary nature of the combinatorial basic operation assembling words together. We will come back to this in the next paragraph.

We have concluded that pure linear sequence is not enough to represent syntax. What shape does the extra dimension give to the representation of syntax? Let us again consider the sequence *John smiles.* The minimal requirement here is that one can take *John* and *smiles* as two separate units and combine them into a larger one. An overwhelming amount of research, stemming in particular from the original work by Richard Kayne, suggests that this minimal binary rule of combination—technically referred to as *Merge*—is not only necessary but sufficient to represent all syntactic combinations. Consider, for example, *John reads these books*; adopting square brackets to indicate the domain of application of Merge, we obtain [John [reads [these books]]]. As we see here, the structure is asymmetric. Such asymmetry is by no means required by meaning or logic—in fact, in some alternative logical notation, *John* and *these books* can be taken as arguments for the biargumental function expressed by the verb *read* (see Chierchia and McConnell-Ginet 2001). Asymmetry in syntax does not follow from anything; it is just an empirical datum, and it follows a long-standing tradition stemming from Aristotle's works on language (see Chomsky 1986; see also Kayne 1994 for an attempt to derive asymmetry and Moro 2000 for further considerations).

Asymmetry is not the only interesting property of syntactic structures, though; another one characterizes syntax, and gives it its special power. Let's first consider these sentences: (a) *John knows this*, (b) *John knows that Mary left*, (c) *This scared John*, and (d) *That Mary left scared John*. We can immediately conclude that the subject of a sentence can be a sentence: here, "this" in (a) refers to the sentence "Mary left" and "this" in (c) refers to the sentence of (b). It is as if a sentence can be "nested" within another sentence. This has a twofold impact on the architecture of syntax: first, it means that part of a structure may have the same architecture as the whole; second, it means that syntax can generate infinitely large structures, as always happens when a system lets an element of a certain type be repeated within an element of the same type. These two distinct characteristics can be referred to as *quasi-autosimilarity* and *recursion*, respectively, and the resulting structures as *nested structures* (to borrow some terms from mathematics). Nesting not only affects sentences; it can easily be detected in all other syntactic structures. In English, for example, it is easy to see that nesting can affect structures that do not involve verbs, such as those consisting of nouns. Consider cases like (a) *pictures*, (b) *pictures of Rome*, (c) *Mary's pictures*, and (d) *Mary's pictures of Rome*. The noun *pictures*, which is technically called the *head* of these particular phrases, can be expanded to the right or left or both by merging other elements containing other nouns, thereby making the group of words assembled in a coherent way bigger and bigger. It is as if these groups had one center (the head), which can expand both to the left (a portion of the phrase called the *specifier*, because it may host elements that can specify the number, intensity, quantity, quality, or act some role expressed by the head, including possession) and to the right (a portion of the phrase called the *complement*, the traditional term that indicates what completes the expression of a verb as *lizards* in *liking lizards*). The group of elements as a whole is technically called a *phrase*—in this instance, a *noun phrase*, because the head is a noun. It is important to note that a phrase behaves as a unit.

This specific way of combining discrete elements ad infinitum via recursion is one of the core aspects of human language and the one never found in any other animal communication code. It has been intuitively captured by many scholars, including Lucretius, Galileo, Descartes, and Von Humboldt, but this is the first time in the history of Western linguistics that its mathematical structure has been captured in such great detail. Interestingly, it can be argued that syntax is not the only code based on discrete infinity that the human mind is able to master, because mathematics and music share this property. This is not to say that this fact allows us to conclude that the three codes completely overlap. For example, as Gennaro Chierchia (2013) has suggested, language is the only system in which lexical items may carry logical instructions—such as those pertaining to set-theoretical relations—but from another point of view one could say that arithmetic is a syntax for a one-word dictionary language. What these three domains do share is the fact that once we allow elements to be combined, there is no logical reason for the computation to have an upper limit. Technically, we say that the set of elements so generated is "unbounded" and that these generative procedures are "open ended." This apparently abstract conclusion corresponds to the strong intuition even children experience that no largest number or longest sentence exists and, more importantly, that there cannot be a proto-syntax—at least in the same sense that there cannot be a proto-arithmetic or an arithmetic confined to a limited upper value (although there can be limited algebraic structures such as the hours of a watch; where adding 4 to 9 we go back to 1). Once you allow 1 to be added to 1 in arithmetic, you allow for the generation of infinite numbers, at least virtually, since the limitations to our memory and our lifespan impose time limitations and we cannot experience infinity.

In this sense, when looking at the structure of a sentence in a human language we may be struck by the strange analogy with snowflakes: minimal components, combined with simple rules that are recursively applied, give birth to geometric patterns of great complexity. The major difference with respect to snowflakes is that sentences must undergo a process of linearization that flattens out the hierarchical bidimensional structure into a linear one, whose words organize what we wish to express from our mouth to another's ears via a cognitive and articulatory "hourglass effect": a linear thread of words moving the contents of one glass bulb (the brain of the speaker) into another (the brain of the hearer); we will come back to this surprising restriction. This is only required by externalization procedures linked to the way our sensorimotor systems are organized. If we were able to speak within a two-dimensional syntactic space rather than the monodimensional space of sound transmission, such a process of linearization would not be needed. Moreover, the resemblance of sentences to snowflakes combined with discrete infinity has a deep impact on the theory of the evolution of language (to which we will return) for at least three reasons. First, any sentence utilizes, in a sense, the entire structure of grammar, just as any arithmetical expression utilizes the entire structure of arithmetic. Second, there can be no proto-syntax since the core notion of syntax by definition involves infinity, and there is no such thing as proto-infinity. Third, there is no room for evolution, but there certainly is for change; snowflakes, after all, do not have a history—they may all be different, but no single one of them is simpler than any other or the ancestor of any other. In this sense, it is language as a whole that is manifested in the structure of each and every sentence, and we can also regard a human language as an upper-dimensional snowflake, or perhaps a snowfall. This has also an interesting meaning when it comes to evolutionary perspectives, of course as far as syntax is concerned. To quote Noam Chomsky, "Language is more like a snowflake than a giraffe's neck. Its specific properties are determined by laws of nature; they have not developed through the accumulation of historical accidents" (see Moro 2012, 2016 for a critical comment on this quotation). Here—as it seems to me— we face one of the most striking and destabilizing paradoxes of nature and surely the one that pertains to us most: a finite object shaped by evolution (the brain) expresses a code that generates infinite discrete structures and that cannot evolve by definition (syntax). Once more, human language reveals itself as the constant scandal of nature.

The impact of recursion on the architecture of human language, when combined with the hypothesis that rules are applied by exploiting the hierarchical structure rather than the linear one, is dramatic. Consider, again, a two-word sentence like *John runs* and the dependency between the two words based, among other things, on the number agreement, which is singular. Similarly, other dependencies are established in sentences like *This fact surprises me*; the number agreement between [this fact] and [surprises me] is also singular. Due to the possibility of nesting, we can have the following sentence: *This fact that John runs surprises me*, where the number dependency of the first sentence is nested within the second. The set of dependencies can

increase progressively, as in *This fact that John whom Mary loves
runs surprises me*. Similar cases can be constructed with words
that do not inflect, such as the pairs *if* and *then,* or *so* and *that*.
Take the following examples: *If Mary plays tennis then she is happy*
or *If John says this then he is a liar*. Or consider even more baroque
entanglements like *If John says that if Mary plays tennis then she
is happy then he is a liar* or *Mary is so smart that she could prove the
Riemann hypothesis in a few hours*. Or even combine both word
pairs and produce this extreme example: *If you say that if this
happens then it is a welcome fact then you must be so angry that I'd
better not meet you.*

One straightforward conclusion we can draw from all this is
the following: An impossible language is one in which dependencies can be rigidly determined by the position of a word in
a linear sequence. This simple statement captures the essential
nature of syntactic rules and simultaneously suggests a precise
and explicit way to define an impossible language. All rules in
all existing and attested languages do in fact respect the hierarchical recursive structure; this phenomenon is sometimes
called *structure dependency*. Interestingly, not all structural phenomena involving hierarchy in human languages are recursive.
Syllable structure is a prototypical example. Simply put, phonol-
ogy describes syllables as being "obligatory" when they have a
nucleus—typically, a vowel—that can be preceded and followed
by a consonant (or group of consonants), which are named *onset*
and *coda*, respectively—such as with the syllable *tan*, formally
represented as [/t/ [/a/ /n/]]. Many phenomena suggest that the
nucleus and the coda form a unit, which we call *rhyme*. There is
no doubt that this is a hierarchical structure, and even less doubt
that this hierarchical scheme is not recursive. It does not make
any sense to have, say, another rhyme nested within a rhyme,

as opposed to a syntactic tree where a sentence may be nested within another sentence. It is recursion, then, rather than hierarchy per se that characterizes the syntax of all and only human languages. Of course, there can be differences in the specific strategies one language (or one person) uses to express ideas—for example, one can prefer to express different sentences (parataxis, as in *It is raining and I want to get an umbrella*) rather than nesting them (hypotaxis, as in *Since it is raining then I want to get an umbrella*)—but nevertheless, there is no doubt that recursion is manifested and pervasive in all human languages. Recursion is the real exclusive fingerprint of human syntax, and syntax is the real exclusive fingerprint of human language. An impossible language is a non-recursive one.

4

THE UNREASONABLE SIEVE

What if every combination of words in a human language was assigned a meaning? We instinctively know that for any given set of words only a very few combinations among its potential strings are acceptable. Many strings of words are ruled out because they convey logical fallacies, such as contradictions or tautologies, or they simply bring together incompatible properties, as in the famous *Colorless green ideas sleep furiously* or the medieval example that alludes to an uninterpretable "triangular circle." Finally, many combinations are just ruled out because they do not contain enough information for the structure to allow for any interpretation, like, say, a string of definite articles. These reasons are already sufficient to negate the possibility that every combination of words can be assigned a meaning. But if we compare languages we can grasp what the advantages are if all combinations of words were possible. Consider, for example, the following two words in English (*Peter, telephones*) and the following two in Italian (*Pietro, telefona*). In English we have only one combination available (*Peter telephones*), whereas in Italian we have two (*Pietro telefona* and *telefona Pietro*)—and the two different sequences in Italian are not completely synonymous. (When one chooses the second, it's the person the speaker wants to focus on as opposed to the first where it's the action). A language in which all combinations of words were possible would be extremely powerful because it could convey much more information—since the number of strings would be much greater, if we assume that synonyms were not significant—and it would have practically no rules to learn (other than Merge). Such a language does not exist, of course. So another way to approach the definition of a possible language is through understanding why not every combination of words is acceptable, apart from the relatively trivial ones we've just excluded for the reasons already mentioned.

One of the examples we considered earlier becomes useful again: *Who do you think Mary wants to describe themselves?* In this sentence, the element *who* plays the role of antecedent of the pronoun *themselves*, but it is not close enough in the sequence to be directly accessible (cf. *Mary wants the boys to describe themselves*). The conclusion was that this element had to be interpreted as also being in a closer position to the pronoun, the same as *the boys* in the latter example: *Who do you think Mary wants ~~who~~ to describe themselves?* This phenomenon is technically called *movement*, as if one element (in this case, *who*) were moved from its original position: which is to say, the position it had occupied before Merge generated the combination of *Who do you think* and *Mary wants to describe themselves*.

Merge has been succinctly defined by Chomsky as "an operation that takes two objects already constructed and forms a new one" (in Moro 2015, xix). The central assumption is that Merge is not limited to just combining new lexical items (which we would call *external Merge*), however, but that it can also copy those that have already been introduced into the computation (which we would call *internal Merge*), as in the case of moving "who" in the example above. Movement thus turns out to be nothing but a case of Merge and, contrary to previous versions of the theory, it does not involve any special syntactic operation. In this sense, the phenomenon of movement is, at least theoretically, expected: a lack of it would be surprising. One interesting thing about the majority of these operations is that the phonological content of all copies but one is deleted. In other words, in our example, only one copy of *who* is pronounced; the other (the barred one) is simply intended at the semantic and syntactic level but is not realized at the phonological level. Moreover, movement is not a special operation: the very fact that it preserves the structure of phrases is not stipulated (as suggested in the influential work by Emonds 1976); it is just a welcome consequence. Understanding what triggers movement and induces deletion remains an ongoing program of research, which I will briefly address at the end of this chapter. For now, the main point is that movement is very useful for understanding impossible languages. A simple case study will illustrate this. Consider the following sentence: *John wants to hire Mary before meeting Hanna*. There are three individuals in the sentence, so we can construct three distinct interrogative structures by exploiting and moving the proper pronoun *who* in order to associate it with John, Mary, and Hanna respectively: (a) *Who wants to hire Mary before meeting Hanna?*, (b) *Who does John want to hire before meeting Hanna?*, and (c) *Who does John want to hire Mary before meeting?* The last sentence is clearly ungrammatical, but there is no reason why this should be so. The type of the information requested by this sentence is by no means absurd; someone is simply asking which person is such that John wants to interview Mary before meeting that person. In fact, there is no logical reason why this sentence should not be grammatical. Nor can it be a matter of complexity or length, because the following sentence is much longer but easier to parse, and is substantially comparable to sentences (a) and (b) when it comes to acceptability: *Who does Peter say that Audrey thinks that Laura hopes that John wants to hire before meeting Hanna?* Now consider these other groups of sentences: (a) *Mary wants to describe a bridge of Rome*, (b) *Of which city does Mary want to describe a bridge?*, (c) *Mary wants to describe a planet on a bridge of Rome*, and (d) *Of which city does Mary want to describe a planet on a bridge?* Movement of the sequence *of which man* is clearly blocked in (d). Again, there seems to be no logical fallacy here. The reason I chose these two examples, though, is that these two apparently unrelated violations of movement can be explained in a unified way. What the two cases have in common is that movement does not take place from a complement of a verb but rather from a phrase that adds information to the sentence but is not strictly necessary, and it is technically called "an adjunct." In fact, one could say *John wants to hire Mary* or *Mary wants to describe a planet* without adding any further information; if the complements were omitted, instead, the sentences would then be ungrammatical: *John wants to hire before meeting Hanna* or *Mary wants to describe on a bridge of Rome.*

This interesting case study reveals at least two distinct and crucial properties of human syntax. First, a sieve exists that filters out several types of dependencies of which—the sieve, that is—we are just not intuitively aware; we don't need instructions to know that *Who does John want to hire Mary before meeting?* is ungrammatical; we know it instinctively. Second, the effects of this sieve cannot be explained by appealing to logical reasoning or common sense; it rather has to rely on formal grammatical notions such as the one of complement or adjunct. Moreover, notice the following fact on the technical side: the status of a verb's complement is not vaguely expressed in syntax; rather, it is encoded configurationally and so are all other grammatical functions, as suggested in Hale and Keyser's fundamental framework (see Hale and Keyser 2002). Thus, the prohibition of movement from a position that is not contained in a complement can be unambiguously expressed in configurational terms. All in all, this case study on the restriction of movement illustrates a domain of research called the *theory of locality*, which aims at capturing the way potential dependencies are filtered out by syntactic principles. The aim of this excursus is to show that the sieve is not based on "rational" or "reasonable" principles—that is, on principles based on reason, logic, or common sense. Rather, it depends on configurational restrictions that fall beyond our immediate intuition and for this very reason they become crucial for the issue at stake here: one can make no appeal to "rational" or "reasonable" principles or to "common sense" when trying to capture the difference between possible versus impossible languages; one must rather explore the restrictions in term of a hierarchical, recursive metric. After all, there is no reason in a snowflake.

Let's take this one step further. The point is that locality doesn't affect movement alone: it can affect many other types of dependencies, such as agreement or pronoun interpretation. Let us consider them separately. Simple cases of locality of agreement are offered by sentences like the following: (a) *John sings very well*, (b) *These friends of John sing very well*, and (c) *This friend of Mary and John sings very well*. Why is (c) ungrammatical? We can easily explain these sentences if we use hierarchical configurations; in fact, it is enough to observe that agreement takes place with the less embedded noun phrase. Notice once more that closeness based on linearity, in this case plane physical contiguity, plays no role whatsoever in the locality conditions. Conditions of locality on pronoun interpretation, although involving very different phenomena with respect to agreement, offer similar results. Consider these sentences: *John thinks that he is tired* and *He thinks that John is tired*. It is only in the latter sentence that the pronoun *he* is unable to refer to *John*: *he* is free to refer to *John* in the former. What is interesting here is that the immediate and simple explanation based on linear order—which implicitly assumes that pronouns can only refer to something that precedes them rather than follows—yields a false prediction. A sentence like *When Mary says that he is lazy, John gets very nervous* is perfectly grammatical: we are able to interpret *John* and *he* as referring to the same person, even if the pronoun precedes the element to which it refers. A hierarchical computation is needed if one is to capture the generalization in a proper way; in this case, it is sufficient to say that a pronoun must not have an antecedent in the minimal sentence it is included in. A simple inspection will confirm this conclusion.

Finding unexpected analogies in empirically distinct domains and tracing them back to a set of few(er) common principles—such as those exemplified here with locality restrictions—is not a sideline goal of linguistic inquiry. In fact, *tracing two empirically distinct domains back to a common principle is the very definition of how one provides any explanation in empirical science.* Interestingly, this methodological point characterizing modern empirical science since its origins in the work of Galileo and Newton was already known in the classical world, even if its theoretical setting was completely different. The prototypical example can be found in the laws of motion. If we relied on our senses alone, without the benefit of wonder and reason, we would be led to false conclusions: that water falls to the ground and fire goes

up. It is difficult not to assume that each is subject to different external forces. This is simply not true, however. There is
only one force attracting matter to the ground—gravity—but an
object may move in the opposite direction if it is immersed in a
denser medium, the same way hot-air balloons rise into the sky.
Lucretius already knew this and depicted this effect vividly in
presenting the structure of the world as made of corpuscles and
in unifying the forces governing their movements:

> Now here is the place, I think, in my account,
> To state this fact: by its own power no form
> of matter can travel upward, or move up;
> and don't let bodies of fire deceive you here,
> because they start and grow with upward movement,
> and gleaming grain grows up, and so do trees,
> yet weight itself will always travel downward.
> We must not think, when fires leap toward the roof
> With flame that hurries to taste of board and beam,
> They do this by themselves pushed by no force.
> It's just like blood that from our body pours
> Pulsing and bubbling up and splashing red.
> See too with what forces water tosses back
> Our boards and beams. The deeper we press them down
> And push on them in a body, and heave and strain,
> The more it blithely throws them up and sends them
> Leaping in air well over half their length.
> Yet we don't doubt, I am sure, that of themselves,
> These things all travel downward through the void.
> So too must flames by force of pressure rise
> Up through the moving atmosphere, although
> Their weight, of itself, would struggle to draw them down.
> (Lucretius 1977, book 2, lines 184–205)

Linguistics and physics share the same characteristics that all
empirical sciences share. What's unique to each field of study,

of course, lies in the changing nature of the principles adopted by each field. There are no a priori criteria unifying the understanding of different phenomena. The idea that languages may be interpreted by laws as in physics, moreover, is by no means a novelty: it was first proposed in the nineteenth century by German comparative linguists, and this proposition even coincided with the emancipation of the discipline of linguistics from philosophical speculations on language (see Morpurgo Davies 1998 and Graffi 2001.

As for the phenomenon we are considering here, the very existence of such an unreasonable sieve, which selects from potential dependencies only those that obey some configurational restrictions, is surprising not just because it is inaccessible to our immediate introspection, but also because it is based on the *only* phenomenon inaccessible to our senses—namely, hierarchy—whereas linear order, which is indeed immediately accessible to our inspection and our senses, is completely irrelevant. Whatever impossible languages are, their properties do not seem to match any obvious computational reasoning based on communication; rather, they obey a different computational reasoning, hidden in the human mind, which demands further exploration yet may lie forever beyond the reach of our understanding.

The exclusion of potential dependencies is not the only active sieve in syntax. In recent years, a lot of effort has been devoted to the derivation of other restrictions that reduce the number of possible structures, and that therefore reduce the number of possible languages. In fact, the *reasons* that trigger movement can be considered another sieve.

One important thing to highlight is that even though, as we have observed, movement is just a case of Merge and not a special syntactic operation, we still need to limit it—that is, we need to find the factors that trigger movement (see Breitbarth and Riemsdijk 2004). If movement were completely free we would have the unwanted consequence that a sentence like *John killed Peter* could end up meaning that Peter killed John after successive free movement of *John* and *killed* over *Peter* stemming from the original basic structure *Peter killed John*. There are two major lines of thinking about what triggers movement: a more traditional one, based on morphological principles called *checking theory* (see Chomsky 1995), and a more recent one, based on symmetry-breaking operations called *dynamic antisymmetry* (Moro 2000, 2009). These two theoretically distinct explanations as to what triggers movement have not yet been unified nor either one disproved and the reason is not obvious: it may be that both approaches can be maintained either because they explain different kinds of movements or because they reflect an inherent dual nature of movement (see the unifying proposals in Chomsky 2013 and Rizzi 2015; see also Moro 2015 for a discussion on the twofold nature of movement) much as it happened in physics with respect to the dual nature of light (Feynman 1985). But what is crucial here is that both the abstractness of these principles and the difficulty in capturing them reveal a very important feature of the architecture of human languages: The class of possible languages is severely restricted at every level, and it is restricted by means of principles that are not reasonable or derived from functional requirements that would facilitate communication.

Still, even if we have no way of accounting for this particular sieve, the independent question remains as to why there exist a sieve in the first place. A plausible hypothesis, albeit a speculative one, is that these grammars have been selected over less restrictive ones in order to fulfill an external requirement imposed by evolution—namely, the fact that the human brain is not always capable of acquiring a language spontaneously during the life of the organism it is developed in. There is a limited time in which this acquisition can take place, roughly within the first four years of life. The sieve—which as we have seen is completely irrational and thus cannot be inferred from experience—offers a way of restricting the field of exploration available to computation and make the brain converge onto at least one language within the limits imposed by biological development. Moreover, it would drastically reduce the possibility of interference and noise and avoid the risk of making them totally unusable. The situation could be compared to the way our retinas perceive electromagnetic waves. Human eyes only see a very limited range of frequencies, from about 400 to 790 terahertz; the retina is like a sieve. If it were sensitive to every frequency, we would find ourselves surrounded by an indistinct fog carrying too much information. For something to be acquired by children and then used throughout their subsequent lives as a source of information, it must be severely restricted. Syntaxes and rainbows are similar in this sense: they emerge from the elaboration carried by those selectional capacities that precedes experience.

Syntax is a difficult field of research and one that like all other sciences requires infinite patience and intuition (and luck, if we keep in mind that, as Louis Pasteur said, "chance favors only the prepared mind"). It's like a Rubik's cube: changing one hypothesis will almost certainly affect another one. It is within this unexpectedly complex scenario that the more we explore the structure of language, the more we realize that not only do impossible languages exist, but they can hardly be considered the "arbitrary, cultural conventions" that Lenneberg (1967) disputed: rather, it appears that they're determined by other factors, including biologically driven restrictions. In other words, the reasons underlying the sieve are not all and necessarily related to rationality or thought. Instead, they appear to be profoundly entangled with the biological structure of the organisms—the neural wiring as well as the other components of the body—in such a way that the notion of an impossible language practically coincides with the notion of an impossible human organism (see Moro 2015 for some further observation of the impossibility of a mutant human with no language).

5

THE BOUNDARIES OF BABEL

How can we further pursue the distinction between possible versus impossible languages? Drawing on what we have observed in the previous chapters on formal grounds, I will now show how this distinction can be explored on empirical, quantitative grounds. The best way to begin is by again referring to Lenneberg's statement—"It is so widely assumed that languages consist of arbitrary, cultural conventions"—and highlighting the alleged conventional nature of human language rules, which many contemporary philosophical interpretations took for granted (and it is worth highlighting the fact that Lenneberg referred to Wittgenstein's followers). Lenneberg's goal when he made that statement was to show that there were sufficient reasons based on clinical observation of pathologies to prove that such an assumption was wrong. The goal of this chapter then is to go even further and show that there is now sufficient neurobiological evidence to support the hypothesis that the distinction between possible and impossible languages is reflected in the way our brain works or, in other words, that the boundaries of Babel are not just arbitrary, cultural conventions, without necessarily referring to pathologies. If the assumption criticized by Lenneberg were right, an attempt to explore the brain's reaction to the distinction between possible and impossible languages would be like exploring whether a specifically yellow traffic light is what causes a person in a car to wait; whereas societal convention could obviously have utilized another color—white, for example—to convey the same instructions. Traffic signals offer a good example of convention and arbitrary association. Our advantage with respect to Lenneberg is twofold: we can rely on a refined notion of syntax (in particular on the notion of recursion) and we have new empirical access to the brain's activity that doesn't require pathology.

In characterizing the contemporary view of language rules, and more specifically syntactic rules, Lenneberg presented three distinct properties: they are allegedly arbitrary, cultural, and conventional. These three conditions, although they may partially overlap, are not necessarily associated with each other: in particular, language could have an arbitrary structure that would be neither cultural nor conventional. One possibility is that the structure of a language emerged spontaneously from chance, from a random set of possibilities that became established through use (see reference to Turing's ideas in Moro 2016). This possibility is particularly interesting when associated with the hypothesis that the human species has a unique origin. Although there may be no agreement as to precisely when our first ancestors emerged within a long history of evolutionary selections, there is a strong consensus that this would have happened in West Africa no later than some 100,000 years ago (see Tattersall 2012, and references cited there). The monogenesis hypothesis could justify the fact that impossible languages exist, but in this case the term *impossible* would be just synonymous with "not yet realized," since a common origin for all languages would make the boundaries of Babel only a temporary limit, and one that could possibly change through cultural or conventional factors.

There is not much of an alternative here: to clarify and possibly answer the issue posed by "Lenneberg's problem," we would need to test whether some of the core properties contributing to the distinction between possible and impossible languages induce differential activity in the brain that cannot be characterized as arbitrary, cultural, and above all, conventional. And to do so we can now turn to what we outlined in earlier chapters and which constitutes one of the major discoveries of modern linguistics: what distinguishes human language from the communication codes of all other species is discrete infinity as carried out by syntax—that is, the capacity of a language to recombine in a recursive fashion a limited set of elements to create an open-ended set of structures (disregarding the impact of the locality conditions imposed on the system). We have seen that one of the consequences of this special design is that no dependency between two words can be established if it is based on the position of a word in a linear sequence (as opposed to the hierarchical structure). Formal linguistics, then, gives us the unique opportunity to test the reaction of the human brain to language rules that do not respect this general principle. There are two preliminary empirical problems to be approached here, which also have ethical aspects. The first issue regards what we can understand about human brains; the second regards what we can understand about syntax, provided that we have access to brain activity. Let's consider them separately.

Testing human brains can be both a trivial matter and a very sophisticated one. We are in fact testing our own brains at this very moment. There are two actors: me, through the act of having written these words, and you, through the act of reading them; our brain is ultimately responsible for decoding what passes through our eyes, and this can easily be proved. For example, if I write this sentence backward your brain will stop the process of decoding: decoding of process the stop will brain your backward sentence this write I if. However, this is only a partial test of brain functioning, and it is what would generally be dubbed a "behavioral test": it can tell us a lot about the way the brain processes information, but it is not immediately useful for our purpose, which is to prove that the boundaries of Babel are not conventional. Nor can we overcome this impasse by testing animals, as is done when exploring other cognitive domains such as vision, since we know that no other animal languages have syntax, let alone a recursive one. Another possible option is found in research on brain damage (i.e., clinical studies). This has indeed been an extremely important source of data, including in Lenneberg's own research. Not only that, but it has been responsible for emancipating studies of the brain from psychology. In fact, we know the exact moment when this began. It was in Paris in 1840, when a young man, Louis Victor Leborgne, arrived at Bicêtre Hospital with an embarrassing and scary symptom. He could only use one word (in fact, just meaningless sequences of the same syllable, *tan*) to express himself, though all his other communication skills, including those involving the use of hands and facial expressions, were intact. Monsieur Leborgne—soon nicknamed "Tan" or "Tantan" by the staff—was hospitalized and, unfortunately, went through a long, painful, and irreversible path to his eventual death in 1861, after experiencing progressive paralysis of his right side. It was the autopsy performed on his body that marked the very moment when our knowledge of the biological properties of language changed. The doctor who conducted the autopsy found a *rammollisement* (softening) of Leborgne's left frontal lobe; this discovery was immediately recognized by the physician as evidence that language not only depended on brain activity, but on a specific portion of the brain. The name of the doctor who performed the autopsy was Paul Broca, and this portion of the brain was subsequently named *Broca's area*. This discovery was immediately accepted by the scientific community—although the discovery was published, curiously enough, in the *Proceedings of the Society of Anthropology*, an institution founded by Broca himself. Medicine has pursued this line of inquiry with great intensity and many discoveries have been made since that epochal one. Interestingly, there was little overlap between the research in medicine and linguistics for at least a century after Broca's discovery. Clinical studies are still a great source of information on human language, but neuroimaging techniques introduced a major change in the neurosciences in the last quarter of the twentieth century (see Whitaker 1998, Bambini 2012, Cappa 2001, and Cappa, Moro, Perani, and Piattelli-Palmarini 2000; see also Domanzki 2013 for a detailed reconstruction of the Tantan story and Poeck for similar cases).

These new technologies allow us to look at some aspects of brain activity in healthy subjects and provide information no one could have dreamed of accessing even in relatively recent stages of the history of science. If we focus just on what is essential for our discussion, we could simply say that neuroimaging techniques give us information on blood perfusion in the brain—that is, on the amount of blood and its components, oxygen in particular for fMRI (technically referred to as the *blood-oxygen-level dependent* or simply *BOLD* signal). This type of information is valuable because it can indirectly allow us to understand whether activity in a certain portion of the brain has increased or decreased. Unfortunately, this method has an intrinsic draw-back. In the absence of pathologies, the brain is always fully active, which means that looking at it while it is performing a certain activity or experiencing a feeling would result in no immediate information. In fact, much of the brain's activity is performed to inhibit activity—for example, if you are sitting in a chair, your brain must send your legs a command to stay still; if you are speaking English and also happen to know Italian, your brain needs to inhibit the grammar you are not using as you read these words on the page. How, then, can neuroimaging techniques be useful? The central "dogma" of neuroimaging, so to speak, is that to get a piece of information you must utilize a comparison: a comparison of two minimally different conditions (the *subtractive method*) or a comparison of different stages of the same condition where one variable is modulated (the *parametric method*). A simple example will clarify this. Suppose you want to know which part of the brain controls the opening and closing of your left hand. One possible approach is to get a subject to undergo two separate neuroimaging sessions, say fMRI: one in which the subject does not move at all (that is, is in a resting condition; see Raichle and Snyder 2007) and one in which the subject moves only his or her hand. We can then compare the two measurements. If the brain turns out to exploit the same amount of energy from its blood flow, the subtraction method would result in a zero; if the brain turns out to exploit a different amount of energy from its blood flow, subtracting would provide either a positive or negative result, but in either case, some sort of concrete figure (see Friston 1997, Price 2012, and Cappa 2012 for extensive reviews).

This just solves one of the problems—that of accessing the brain's contents without having to wait for a local lesion to appear or, worse, for an autopsy to occur—but it introduces a second problem, and a major one at that. Comparison, as we saw, is key to obtaining information through neuroimaging techniques, but syntax can neither be suspended when examining linguistic activity nor modulated in order to have a subject perform a task that is more syntactic than another. Syntax, from this point of view, is like arithmetic: it's either present as a whole or it isn't, and if it isn't present, then neither is human language. Once you get from n to $n + 1$, you already have access to infinity, and once you are able to put together two words to form a sentence, you already possess syntax. How can neuroimaging be used to explore syntax if we need to get through some form of comparison, then? This is the essential preliminary question for any experiment on impossible languages: if one cannot isolate syntax and disentangle it from other grammatical components, one cannot address the question of whether impossible languages exist. Two independent experiments addressed this very issue and converged toward the same goal (Embick et al. 2000; Moro et al. 2001). In both cases, the working principle was that, since one cannot suspend syntax, at least one can explore whether the brain reacts to different kinds of errors, including purely syntactic ones. So, for example, working from a sentence like *The boy runs*, one can test such variants as *The boys runs*, *The boy rtnss*, or *Boy runs the*. Some additional precautions were taken in the second experiment cited here to avoid confounding semantic competence with a syntactic error. This requires a slight detour from the major line of reasoning. Suppose one takes a sentence like *A tiger has killed a hen* and destroys its syntax so as to produce *A has tiger hen killed*. The knowledge of the meaning of the words *tiger*, *hen*, and *kill*, sometimes referred to as our "encyclopedic knowledge," is probably sufficient to restore the original intended meaning, but if the sentence is *A hen has killed a snake,* since either of these animals can be presumed capable of killing the other, a disrupted syntax may give a spurious result. To solve this, it was proposed to test sentences composed of pseudowords: words that sound like possible words in a certain language but whose roots have not been assigned any meaning. By combining pseudowords with function words we can construct pseudosentences like *The gulk ganfles the brals.* No truth value can be assigned to this sentence—that is, no one can say it is true or false—and thus it constitutes a sentence with syntax but no semantics. Comparing this sentence with modified examples like *The gulks ganfles the brals, The gulkzrts ganfle the brals,* or *Gulk the ganfles brals the* would provide the ingredients we need for comparison so as to perform neuroimaging data subtraction. It goes without saying, first, that one has to invent and test a good number of different sentences, even before turning to neuroimaging technology, so as to avoid evoking any inadvertent meanings; second, there is no a priori full guarantee that this method will work. It will work only if it allows us to obtain some data; otherwise, it might well be that the number of variables involved is just too vast for this simple test to tell us something about syntax and the brain.

What turned out to be the case, however, is that both experiments produced the same results; the second experiment, the one excluding semantic interference, just allowed for a more refined analysis in that it showed that syntactic errors employ a twofold system that involves a deep component of Broca's area (in particular, the pars triangularis of that area) and the left nucleus caudatus (an inner portion of the brain belonging to the basal ganglia system). Interestingly, these results, including the ones involving the left nucleus caudatus, have been confirmed by a third independent experiment carried out some ten years later, which assumed the same empirical strategy and extended it to address inferential errors (Monti, Parsons, and Osherson 2009). These preliminary results allow us to address the major empirical issue concerning the nature of the boundaries of Babel; we can now safely assume that syntax is autonomously represented in the brain. One last important remark needs to be made here, though. The very fact that we can see the activation of a certain area in the brain in relation to a certain task being performed does not mean that that task is exhaustively computed in that area. A certain area, like Broca's area, can play the role of a hub airport in the air traffic system and be activated because some airline path passes through there. It could also be that—as is the case with hub airports—Broca's area is just the place where important paths intersect. Sometimes major hubs are not close to major cities; they just host convenient ways of getting to major cities. Similar considerations can affect our understanding of the brain when we apply neuroimaging techniques. Moreover, the complexity of the brain overwhelmingly exceeds the scarce information one can get from blood flow. To get an idea of the dimensions of this complexity, the following comparison may work: "If we counted one synapse per second, we would not finish counting for 32 million years. If we considered the number of possible neural circuits, we would be dealing with hyperastronomical numbers: 10 followed by at least a million zeros. (There are 10 followed by 79 zeros, give or take a few, of particles in the known universe.)" (Edelman and Tononi 2000, 38). Also, the electrochemical communication between neurons, unlike most artificial electric circuits, is not necessarily either on or off, but instead it may range over all the values between the two extremes (Kandell 2012). This fact greatly increases the information-encoding possibilities: neurons do not work like electric switches, but can assume a potentially infinite number of intermediate values, the only restriction being the molecular limits of cell structure at the level of quantum theory interactions (see Penrose 1989 for a critical discussion of this issue). It is, again, like looking at the Manhattan skyline during the night but with a further complication; not only does the fact that light may or may not come from any single window make a difference, but so does the question of whether that light is regulated by a dimmer between the two possible opposed values of maximum and minimum. The number of combinations is virtually infinite, yet the Manhattan skyline is the same object, from a naive observer's point of view. Reconstructing brain activity from one unique variable related to blood flow and oxygenation appears even more ambitious than trying to construct maps of the Earth's major cities from observing the flow of passengers in airports, but it's not completely hopeless. While we may not be able to obtain every bit of information, and perhaps not even the most interesting bits of information, through neuroimaging techniques, these techniques can still provide us with some conclusive data regarding the nature of impossible languages.

We now have all the tools we need: a formal theory of possible languages, a technology that allows us to see where the brain is active, and the knowledge that syntax activates an integrated network of areas in the brain. I had an opportunity to work with three different teams to test the core hypothesis that the brain does not consider rules based on linear order as part of the language system (see Moro 2013 for the collection of papers illustrating these results). Though our approach varied slightly, out strategy was essentially the same each time: to test the acquisition and processing of syntactic rules belonging to either linear or hierarchical dependency rule systems. If the brain reacted in different ways, then we would have to dismiss the assumption that the distinction between impossible and possible languages is arbitrary, cultural, and conventional. Let us consider the types of rules exploited and the setup of the experiments. I will limit the description to one experiment, namely Musso et al. (2003); for further technical details see again Moro 2013 or the less technical illustration in Moro 2015. A group of twelve subjects who had been exposed to only one language over a lifetime was selected. They all came from the former East Germany, where, as in most dictatorial environments, people were forced to speak only the one language approved by the government. In this case, the twelve people spoke German. They were taught a version of micro-Italian including only a limited set of nouns, verbs, and some basic function words such as articles, particles, auxiliaries, negations, and, of course, some syntactic rules. Some of these rules were actual rules of Italian—for example, we taught them that in Italian one can form a sentence without expressing its subject, unlike German (and French and English, among others); they were also taught how to construct an embedded sentence, a construction very different from the matrix sentence in German but not different from the Italian matrix sentence. Then they were taught "impossible rules": rules with rigid dependencies based on the position of words in the linear sequence, running against the specific recursive structure implemented in human syntax. What follows are three examples. The first rule was for constructing a negative sentence and required specific positioning within the sentence: insert the word *no* as the fourth word of the string. The second rule involved the two extremes of the string constituting a sentence and specified that the first article of the string obligatorily agrees with the last noun. The third rule instead involved the string as a whole and required that all the words of a sentence invert their linear order in any interrogative *yes-no* question. The subjects, of course, were not aware that these rules were based on two different major types (recursive and linear) and started to learn how to process the rules. The experiment consisted of testing the brain's reaction at different stages in the process of learning. A different group of eleven subjects was also given a micro-Japanese course, after one reviewer asked us to address the fact that German and Italian both belonged to the Indo-European language family, which introduced a variable that could contaminate the results. A similar set of linguistic stimuli was then designed and the subjects were tested during the learning process.

The results obtained by measuring the BOLD signal with an fMRI were very clear. We concentrated on the activity in Broca's area since we knew from the previous experiment exploring selective errors with pseudowords that that area was selectively involved in other components of syntactic processing; if we didn't get any results we could have checked other elements in the network dedicated to syntax. This activity was checked against the subjects' ability to master the new rules in the new languages. To avoid complications with sound and with different writing systems, the new micro-language grammars and the exercises were all written in the Latin alphabet. All in all, the experiment showed that the amount of blood in Broca's area *augmented* when the subjects increased their ability to apply rules based on recursive architecture, whereas it *diminished* when the subjects increased their ability to apply rules based on linear order, regardless of which micro-language was involved. I have also reproduced similar experiments while working with different teams and center or research by varying the type of stimuli

(Tettamanti et al. 2002, Tettamanti et al 2008; and also published in Moro 2013). What remained constant was the contrast between recursive (or structure-dependent) versus linear rules, assumed to be at the basis of the distinction between possible and impossible languages. In the former case, the subject learned a language made of pseudo-words, to avoid potential confounding effects of lexical semantics, and received no explicit information about the rules: they had to figure the rules out for themselves. In the latter case, the string of elements was not interpretable at the phonological or ideographic level: they just looked as if they were the mysterious symbols of an unknown language. The rules, in this case, were based on the color, shape, or dimension of the symbols.

All three experiments produced the same converging results: the brain distinguished the two types of rules—recursive versus linear (i.e., non-recursive rules)—because Broca's area reacted differently to them. When the subjects improved their ability to deal with recursive rules, Broca's area progressively increased its activation; when the subjects improved their ability to deal with linear (non-recursive) rules, the activity in Broca's area was progressively inhibited. The conclusion was inescapable: since we lack control over the brain's circuit activity—and in fact we lack even any intuitive awareness of it—it is hardly possible to claim that language rules consist of "arbitrary, and cultural conventions": if they did, the very same network dedicated to language should have been activated, and no difference in Broca's area should have manifested. Lenneberg's polemic claim in favor of a biological interpretation of language structure was thus bolstered. The behavioral tests in the three experiments also converged to show that the subjects did not find one family of rules more difficult than the other. The number of errors was the same and so were their reaction times (apart from one small deviation in one case in favor of possible rules in Musso et al. 2003), which excluded the possibility that difference in brain activation is due to memory capacity or any other, possibly unknown computational effort (see Chesi and Moro 2014 for a detailed computational analysis of the rules exploited in the experiment in terms of automata—that is, abstract automatic machines).

Of course, as always happens in empirical science, one answer raises many new questions. One immediate question, for example, is: what circuits in the brain computed the impossible languages? Any answer to this question does not challenge the result but raises an important issue. The experiments converged in getting three types of results: first, the involvement of the right hemisphere; second, nonspecific activation for impossible rules; and third, the partial involvement of networks that are generally active when computing solutions to problems. Another question raised concerns the activation of Broca's area for non-linguistic tasks, such as motor or computational tasks, as in music or mathematics. In the latter two domains the question becomes very challenging, since it is clear that the type of code exploited in these two domains has recursive properties. Note that in a modular view of brain functioning it should hardly be surprising that a certain area is active during two or more different tasks. This is what one would expect when adopting such a point of view, because this is precisely what defines modular systems. Modularity is not necessarily linked to a biological system. One can think of other cases. Suppose, for example, you need to organize a large kitchen in a restaurant. There are two possible approaches: you could hire as many cooks as there are dishes on the menu, or you could distribute among a group of cooks special tasks, none of which in itself results in a finished dish.

One cook may be a specialist in salting water, another in whipping cream, and so on. In this way a cook may work with other cooks on different dishes. The advantages of the second system are that there are no redundancies and you need fewer cooks than the number of dishes you are offering at the restaurant; the disadvantages are, again, that there are no redundancies, so if a cook calls in sick, there will be more than one dish you won't be able to offer. Nature seems to choose the second option, perhaps relying on the fact that—as when you organize a kitchen—cooks can learn to perform other tasks in an emergency. Broca's area may also be involved in different tasks, depending on what other components are active; the caudate nucleus, for example, gets selectively activated for syntactic computations.

Nevertheless, it should be emphasized that this experiment does not reveal the precise role Broca's area plays in language processing and a more finely grained inquiry into this question has only recently been undertaken (see for example the work by Angela Friederici in Friederici et al. 2006 and Zaccarella and Friederici 2015); it does show, however, that the distinction between possible and impossible languages is embodied in the brain. This not only meets the potential objections Lenneberg was contrasting as to the alleged arbitrary, cultural, and conventional character of human languages and, as we noted, resolves the polemical debate in favor of a biological interpretation of language structure. From a different point of view, it also dismantles a popular conviction that languages are forms of software running on a passive hardware base; if anything, they are the expression of the hardware activity, as if flesh became language, *logos*. The distinction between possible and impossible languages, obtained from comparing formal structures of languages, has finally been solidly anchored to a neurobiological phenomenon.

6

INTERMEZZO OR WHAT EXISTS
WHEN A TREE EXISTS?

It is common practice to say that linguists provide abstract representations of language structures, whereas neurobiologists provide concrete ones. I think this is a mere prejudice, or rather, a misleading conclusion. In this chapter I address the ontology of syntactic trees and try to offer some clues to allow us to properly address the following question: What exists when a tree exists? The reason this question must be raised here is that we do not want to end up with two variants of the distinction between possible and impossible languages, a concrete one and an abstract one. We have already seen that fruitful comparisons can be made between the domains of theoretical syntax and neurobiology, but the issue concerning the alleged abstractness of syntactic representations still remains.

There is, of course, a cognitive bias: a syntactic representation of a linguistic structure typically rendered with a binary graph called a *tree* is different from a piece of human brain tissue in which the syntactic processing takes place.

As a first approximation, a tree is a set of descending segments (called "branches"), each segment departing from a single point (called a "node"). Where the segment ends there can be a word (or a part of a word; i.e., a morpheme) or another pair of

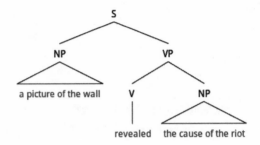

FIGURE 6.1

A syntactic tree. S = Sentence, NP = Noun Phrase, VP = Verb Phrase, V = Verb.

segments reproducing the same pattern as the first two. Each node is normally labeled indicating the syntactic object produced at that level of computation (see Kayne 1994, Moro 2000, 2013, and Chomsky 2013 for a comprehensive framework).

The question, however, is whether any other empirical science provides less "abstract" representations of the way their object of inquiry is structured. Consider physics, for example. We have been used to the so-called solar representation of the structure of atoms (technically known as the *Rutherford-Bohr model* since the early twentieth century). Everyone is able to recognize it: it looks like a micro–solar system, with the atomic nucleus at the center and the electrons moving around it in circular orbits much as planets are represented to orbit around the sun. We now know that quantum physics has radically changed this representation, since it has demonstrated that our notions of particle and wave are entangled in a complex way (see again Feynman 1985). We could then legitimately ask: What exists when we assume that the Rutherford-Bohr model of the atom exists? The answer can be easily found even in elementary handbooks of physics: this representation is an attempt to capture the results of experiments in a unified, synthetic (and telling) way. One of the central experiments was the one in which alpha particles passed through gold foil. By calculating the number of particles passing through, the number of those deflecting, and their trajectory, a physicist could conclude that the atom had a nucleus that was positively charged (protons, along with neutral neutrons) and that negative charges (electrons) orbited at deter- mined fixed distances from the nucleus. When a solar model of an atom exists, it means that there exists a set of results from an experiment on otherwise unknown matter.

It is not, strictly speaking, necessary to interact with matter at every step of a research process to understand how the world functions and to provide an empirically valid model. Consider, for example, the prototypical case of the representation of fall- ing bodies on the Earth: a representation derived from Galileo's groundbreaking work that led to the development of modern science. The traditional pictorial representation of this experi- ment has Galileo climbing up the Leaning Tower of Pisa and dropping stones of different weights while calculating the time they took to reach the ground. As in the case of physics, we had someone probing the properties of the world and registering the results. In fact, this never happened, as Galileo explicitly admit- ted. In an interesting passage from his *Discorsi e dimostrazioni matematiche intorno a due nuove scienze* [*Dialogues Concerning Two New Sciences*] ([1638] 1954), two characters, Salviati and Simpli- cio, have the following exchange:

SALVIATI: If then we take two bodies whose natural speeds are different, it is clear that on uniting the two, the more rapid one will be partly retarded by the slower, and the slower will be somewhat hastened by the swifter. Do you not agree with me in this opinion?

SIMPLICIO: You are unquestionably right.

SALVIATI: But if this is true, and if a large stone moves with a speed of, say, eight while a smaller moves with a speed of four, then when they are united, the system will move with a speed less than eight; but the two stones when tied together make a stone larger than that which before moved with a speed of eight. Hence the heavier body moves with less speed than the lighter; an effect which is contrary to your supposition. Thus you see how, from your assumption that the heavier body moves more rapidly than the lighter one, I infer that the heavier body moves more slowly.

The result is surprising. The very fact that all bodies fall to the ground in the same length of time when falling the same distance (absent differences in air resistance based on the shape of the bodies) is derived from pure reasoning: there was no climbing of any tower, and any different conclusion would just result in a contradiction. This is the first example of a "thought experiment," as it came to be known in the twentieth century. The explanation was fully empirical, of course, in the sense that it was based on the hypothesis that what matters is the force pulling the bodies—the gravity generated by the mass of the Earth is so great that the difference between the two bodies' mass does not matter—but the deduction that led to it did not need experiments: reasoning was sufficient.

In physics (though analogous cases can easily be found in any other empirical science) a representation is the synthetic result of either an experiment performed on physical matter or an experiment performed in the mind through reasoning about what we already know (from previous experience). The question now is whether the representation of syntactic relations rendered by trees (or any isomorphic graph) has any experimental counterpart. One has to be careful here. There is one sense in which this representation is already granted by an experiment. Suppose for a moment that we don't know syntax contains a binary compositional rule (Merge) and we have a sequence like the following: *Mary met John*. This sequence would be compatible with three possible outputs, two asymmetric ones and a flat one: (i) [*Mary met John*], (ii) [*Mary* [*met John*]], and (iii) [[*Mary met*] *John*]. Suppose now that a linguist analyzes a sentence like the following: *Mary met John and Peter did too*. The interpretation of this sentence is easy: Mary met John and Peter met John too. Thus, *did* stands for the sequence *met John*; the representation in (ii) qualifies as the obvious candidate to represent the sentence, because it groups the verb and the object noun phrase together. It would not make much sense to consider *met* as connected with *Mary* or as being isolated, as in the flat structure. Note that we have independent evidence to assume that only elements that can be grouped together can also be replaced by single words, much as in the case of pronouns: when we say *The author of the Divine Comedy thinks that he is a great poet* we know that *he* stands for *The author of the Divine Comedy* rather than, say, *The author of*; and notice that there is independent evidence that we can group together *The author of the Divine Comedy* in that sentence since we can, for example, say *It's the author of the Divine Comedy that thinks that he is a great poet* but not *It's the author of that the Divine Comedy thinks that he is a great poet*. What makes (ii) coherent and (iii) not is that no single word can replace [*Mary met*] whereas *did* is able to replace [met John]. This is already a sufficient argument for assuming that the asymmetric representation is empirically justified; why else would things be like this? One would be hard pressed to invoke pure chance as a justification for all of these facts.

In a sense, then, a tree is to language what the solar model of the atom is to matter. What differs is the empirical domain: in physics, the empirical domain is matter; in linguistics, the empirical domain is the reaction of a native speaker.

Nevertheless, a different, probably more critical question can be posed, which is whether the asymmetric representation of a syntactic tree can be used as an empirical argument and can be taken as the synthetic representation of the way the brain works *from a neuropsychological point of view* rather than a behavioral one. Neuroimaging techniques have allowed us to establish empirical arguments that link the representation of a syntactic tree to the neuropsychological processes in the brain. I will describe two of them here.

The first experiment (Abutalebi et al. 2007) involved bilingual subjects; its central goal was to explore the cost to a bilingual subject of suddenly switching from one language to another, as well as to calculate the neural cost of the auditory perception of language. Parallel to that, the investigation explored whether the brain reacted differently if the switch took place between the specifier and the head or the head and the complement. The results demonstrated that such was indeed the case: the brain's reactions were different. Moreover, the first type of switch correlated with the activation of regions typically activated for morphosyntactic processes, such as agreement, whereas the second type of switch correlated with the activation of areas typically activated for lexicosemantic processes. The second experiment (Pallier, Devauchelle, and Dehaene 2011) tried to support the idea that syntax yields a hierarchical structure. The investigators first measured the brain's reaction to a list of words that were not syntactically connected, then they measured its response to lists of progressively connected group of words; first lists of two by two words, then three by three, then four by four, the six by six, up to sentences of twelve words each. The hypothesis was that the neural assembly that encodes a constituent grows with the number of words and this was empirically confirmed. Crucially, the same incremental activity proportional to the number of words combined hierarchically was observed whether true words were used or pseudo-words (such as those used in a previously cited case *The gulk ganfeld the brals*). This showed that the activity seen there was a real syntactic activity rather than one based on meaning, and thus indirectly supported the idea that syntactic trees are genuinely independent constructs.

The findings from these two experiments not only matched those exploring the reactions of native speakers from a behavioral point of view, but they provided different kinds of empirical support in favor of the representation embodied by the tree representation and, based on the first experiment reported here, an asymmetric representation. Indeed, a syntactic tree can be considered on par with the synthetic representation of probes on matter, with the important distinction being that the matter in question is not only organic matter but is part of a living organism. But all in all, one can safely conclude that what exists when a tree representation is adopted has the same ontological status as what exists when a solar model of the atom is adopted.

The experiments exploring the brain's ability to process syntactic representations that we have examined so far all have one characteristic in common: the stimuli are given through ordered elements or through elements that contain an error and the subjects are asked to react to them, either by producing different ordered elements on the basis of some rule or by fixing the error. A recent experiment explored the opposite view (Albertini, Tettamanti, and Moro 2015). Subjects were asked to turn well-formed sequences of words into a disordered structure. The participants read printed sentences or noun phrases aloud, one at a time. All the stimuli contained six words, arranged in different constituent structures. Immediately after one stimulus was read, the sentence was hidden. The subjects were then instructed to repeat the same words as in the stimulus but in a different and completely arbitrary order of their own choosing; they were given no constraints as to how they wished to execute this task, nor offered any examples or hints. The results showed that the subjects could not completely get rid of the underlying phrase structure, albeit unconsciously. Although prompted to recombine words at random, the subjects consistently produced new combinations of specific patterns that were dependent on the type of phrase structure they had been given. Moreover, the irrelevance of lexicosemantic biases, such as the fact that a verb like *burn* does not select abstract complements, and of the frequency of word *co-occurrence* in texts was demonstrated by comparing stimuli formed by actual words with stimuli formed by pseudowords. This experiment provided a novel type of evidence in favor of the psychological reality of phrases in adult grammars. Possible structures, once they are implanted, are very resilient and cannot easily be altered, even voluntarily.

Finally, when considering the nature of syntactic representation, it's worth highlighting the role it plays in linguistics. Syntactic representation is important for at least two reasons. One—the simplest reason—has to do with its synthetic force: a tree representation captures a set of properties in a formal way that would otherwise require a much longer description. In this case, for example, the binary ramification adopted in trees represents the fundamental property of Merge (i.e., that of combining two elements at a time). Another reason syntactic representation is important is its deductive-heuristic value. As in other disciplines, especially physics, formal representation may suggest generalizations that would otherwise go unnoticed or would at least be much harder to see; the similarities discovered by Charles Augustin Coulomb in 1788 between the electrostatic interactions of electrically charged particles and the gravitational interactions of masses emerged in a natural way from the format of the formulas used to capture them (see Feynman 1965). Similar heuristic paths can be found in linguistics: for example, in the search for general locality conditions of movement, the comparison between the restrictions imposed on noun phrases with those imposed on sentences allowed some very powerful and general restrictions to be discovered (Roberts 1988, Graffi 2000); we saw a simple case when we discussed the unified treatment for two apparently distinct cases such as *Who does John want to hire Mary before meeting* and *Of which city does Mary want to describe a planet on a bridge?*

Of course, tree representations do not exhaustively capture the properties that constitute the domain of syntax. This is not due to any flaw in the representations, which may vary a lot, but rather to the fact that any analysis must choose from the whole set of properties necessarily "idealizing" the field, as assumed in the "Galilean style of research" that is generally traced back to Edmund Husserl. One typical case pertains to the development of a fine-grained analysis of the building blocks of clause structure. In the mid-1980s, sentences were considered bipartite: a nucleus where predication relations are established and a periphery that is normally exploited to produce interrogative sentences or special locutionary forces, as in imperatives, and so on. In the late 1980s, two independent studies (Moro 1988; Pollock 1989) suggested that the nucleus should be split into independent domains; a similar process of head splitting has characterized part of the nucleus, the verb-phrase domain (as proposed in Larson 1988 and Hale and Keyser 2002) and the left periphery of the sentence (proposed in Rizzi 1997). The result is that the tripartite structure of the clauses—argument, predication, and illocutionary, moving from the core toward the periphery—was seen as if put under a microscope. The result was that many phenomena that had previously been ignored or unexplained—such as the syntax of copular sentences, the syntax of finite verbs, or focus/topic constructions—were discovered and accounted for by means of a principle-based framework, which evolved into the so-called cartographic project (Cinque and Rizzi 2008). This is not to say that these new projections need to be supported by neuropsychological research; rather, the neurophysiological representations may need rethinking.

An extreme ontology based on explanatory processes could be assumed according to which *something exists to the extent that it plays a role in an explanation*, in all empirical sciences. The overall consequences of this view (as well as the possible contradictions) are hard to determine at this point in time, but what matters here is that impossible languages are no less real than possible ones, since they are needed to describe and understand the faculty of language and all phenomena related to it.

7

TOWARD THE SOURCE OF ORDER

Where does the order in language structure come from? Is it
inherited from some other structure? Is it shared with some
other structure or is rather a singularity in the observable structures of the world? Is it determined by some properties of its
minimal constituents, as an emergent entity? These questions lie
at the very heart of the history of Western thought. In fact, one
could regard this very inquiry as the common thread connect-
ing all moments in the history of human thought on the nature
of language since Plato's first dialogues, if we were to limit our
observations to Western thought. Modern technology may have
allowed us to reformulate these questions in terms of physical
properties, but the interest in the structural analogy between
language and the world has always permeated the field of linguistics as well as the philosophy of language and philosophy in
a broader sense. Clearly, the discovery of any analogy between
language and any other structure is central to the question of
impossible languages; indeed, if any such analogy exists, it may
be responsible for the source of order constituting the boundaries of Babel as I have described them in the previous chapters.
I shall reflect on this source of order in two steps: first, I will
consider the analogy issue based on considerations pertaining to

motor control; second, I will explore a novel perspective on language structure available through a new technology that allows us to access neurophysiological data directly from the cortex.

The idea that language inherits its properties from the structure of the world has been taken seriously by many scholars, but probably never as much as by the philosophers known as *speculative grammarians* or *Modistae* ("Modists") who flourished in the fourteenth century, especially after the systematization of this thinking by Thomas of Erfurt in his *De modis significandi seu Gramatica speculativa* (Hall 1972). Indeed, aside from any philosophical explanation, the idea that language inherits its structure from something else is the product of a widespread attitude that characterizes human cognition and reasoning. Suppose you see a shadow on the wall that is moving in a coherent way. Instinctively, you turn your head back to identify the source of that image. We do have parsimonious minds when it comes to evaluating the source of a structure; a similar attitude does characterize many reflections on language, as those of the Modistae. The Modistae owed their name to the *theory of modes*, a tripartite view of the world: modes of being (*modi essendi*), modes of understanding (*modi intelligendi*), and modes of signifying (*modi significandi*). Anchoring ontology to perception and expression—another way to dub these three *modi*—leads to a straightforward result: that of reducing the number of possible languages to one, given the restriction imposed on the ontology, i.e., given that there is only one way for objects to exist (at least in the Aristotelian tradition adopted by the Modistae). Actually, the best synthesis of this view preceded Erfurt's treatise and was formulated by Doctor Mirabilis, the Franciscan philosopher Roger Bacon. In his words: "Grammatica una et eadem est secundum substantiam in omnibus linguis, licet accidentaliter

varietur" ("Grammar is one and the same in all languages in virtue of its substance, even if there can be accidental variations").
This claim is sharp and extreme: there is only one possible language because the grammar underpinning languages can only
have one substance (which is what ontology is about). No one
would accept such a claim today, despite its obvious similarities to neurobiological findings, but the temptation to construct
analogies between language structure and the external world is
still strong.

Beside historical notes, the question to be posed here is:
What else do humans do that may be similar to syntax? Indeed,
there are human cognitive competencies that share a lot with
syntax, such as mathematics and music, and it's worth briefly
considering them. In both of these cases we have a formal system generating an open-ended (i.e., infinite) set of structures by
combining discrete elements in a recursive way; in other words,
discrete infinity is shared by these three domains. There have
been many efforts to explore the similarities between syntax and
music. David Pesetsky, for example, elaborating on the fundamental work by Lerdahl and Jackendoff (1983), has suggested
that music syntax is very similar to language syntax (Katz and
Pesetsky 2011). What is lacking in music, though, is the presence of logical instructions, which characterize the lexicon of
all human languages, as manifested, for example, in function
words (see again Chierchia 2013). Neuroimaging studies have
shown the involvement of Broca's area in music (see Patel 2003
and Sammler et al. 2013 among others), but the data does not
yet allow for an integrated theory of these cognitive capacities.
Technological problems aside, what is lacking is a shared set of
empirical questions. The linear nature of music signals is a complex matter, though: music is indeed linear when it comes to

melody, but it's not when it's a question of harmony. Harmony is a very interesting phenomenon: it includes the possibility of two independent and different melodies being simultaneously processed as they happen (what in music is referred to as *counterpoint*) and are thereby able to convey a unique meaningful and complex content; this can be done with a single instrument as in Bach's "Goldberg Variations" (BVW 988) played on a harpsichord or by a large ensemble, as in a symphonic orchestra.

Language is different, though, because the human mind is just not able to process two simultaneous sentences: in contrast to music, there are no multiple-channel meaningful and complex structures by which two or more sentences can be simultaneously interpreted. In language, there is no possibility for a harmony in which two sentences could be meshed together; we are stuck with melodies: verbal symphonies that yield a comprehensive and meaningful semantic content by composing the meaning of each and every word of simultaneously uttered sentences are impossible.

Mathematics also has other characteristics when it comes to its cognitive status (see Dehaene 1999 and references cited there): it can be argued that mathematical computation and basic arithmetic requires linearity, but their notation can certainly be bidimensional, as in calculus or linear algebra matrices (which, incidentally, make mathematical notation more similar to ideographs than to alphabetical expressions). But in regards to linearity in mathematics, we do exploit language resources when we make calculations; $5 + 2 = 7$ can in fact translate into the sentence *five and two make(s) seven,* a sentence that includes the transitive verb *make*. These linguistic expressions in a language one could call "Calculese" manifest unexpected properties; for example, although the same mathematic expression $5 + 2 = 7$

can be equivalently written as $7 = 5 + 2$, in Calculese the corresponding mirror sentence *seven makes five and two* does not make sense, the same way that—if we utilize a non-mathematical subject and object—saying that *a cook makes a pizza* makes sense but *a pizza makes a cook* doesn't. The situation would of course be different had we used a copula (a linking verb) instead of the transitive verb *make*: for example, *five plus two is seven* will still make sense as *seven is five plus two* (the same way that *a pizza is a cook* still makes sense as a sentence when rendered as *a cook is a pizza*). But copular constructions are also special cases in English (Moro 1997, 2010) that often allow for specular sentences to be grammatical. Prima facie, Calculese appears to have its own grammar that has yet to be discovered, although it largely overlaps with natural language grammar—certainly when it comes to linearization—and it seems to include the equivalents of subjects, verbs, complements, and the like. But as in the case of music, if one can claim that there are common aspects underlying mathematics and syntax, and crucially so in their use of recursive structures, the two domains are clearly too different to simply assume that they coincide on a structural level.

Outside the domains of music and mathematics, there have been some recent proposals suggesting a link between syntax and other cognitive domains and which identify at least two distinct lines of research. One line of research involves evolutionary theories of language that consider language to be the result of an emancipation from gesture and motor control. Michael Corballis's proposal is best synthesized by the title of his book, *From Hand to Mouth: The Origins of Language* (Corballis 2003). The other line of research involves theories—such as the position taken by Pulvermüller and Fadiga (2010)—that are based on the hypothesis that a sequence of actions adopts the same syntax to be found in a sequence of words. Both of these theories explicitly rely on an interpretation (which we might better refer to as an "extension") of Giacomo Rizzolatti's mirror neuron theory. This theory, based on experiments testing macaque monkeys, claims that when we see someone intentionally performing a meaningful action or hear a sound typically associated with a well-identified motor action, or we hear or read words describing a motor action, the same motor network is activated in our brain as if we were intentionally performing the same meaningful action (Rizzolatti et al. 2002, and Tettamanti et al. 2005). I have elsewhere argued against both of these "reductionist" theories (see Moro 2013, 2015), as they overlook the fact that the recursive hierarchical structure I have described in previous chapters is something that is invisible to movement perception, and therefore to any alleged mirror neuron interpretation. It is possible that my criticism of these theories may ultimately prove to be incorrect, but it is worth noting that proving these reductionist views to be wrong would be the better outcome overall. Otherwise, understanding the singularity of human language "among all other animals" (to quote Descartes) would be practically insurmountable, since gestures and action planning are surely shared by all primates (and, to different degrees, other animals) in a very sophisticated way.

What we need in order to approach the analogy issue, though, is not just a change of perspective. Above all, we need to realize that neuroimaging techniques, even if they provide the kind of crucial data I described in the previous chapter, are not well suited to capturing this different level of analysis. Neuroimaging lets us know only *where* neuronal operations take place; we need a different sort of technique if we are to explore *what* neurons communicate to each other when our brains process linguistic structures. We will address this turning point in technique in the next chapter. There is an interesting historical aspect to understanding what the source of order is for human language structure, though, which we should outline first.

The question regarding the origin of the source of order in human language can be approached *within* language, which is to say, without necessarily referring to other cognitive domains. It appears that the complex scenario described by the history of Western linguistics can be reduced to two extreme points of view: at one extreme, order is seen as emerging from an unformed, infinite magma through the spontaneous development of symmetric analogical relations; at the other extreme, order would be formed by unforeseen and unforeseeable fractures in an immense lattice of symmetric regularities, where everything would otherwise be inert since it would everywhere be the same. These two points of view were in fact canonized within the philological debate in the Hellenistic period and they have continued to be held ever since: these two formative principles were called *analogy* and *anomaly*, respectively, and were adopted by the Alexandrian school in Egypt and the Pergamon school in Asia Minor (Lesky 1971). There is a further complication to this scenario, though: for a structure to be recognized, it is not necessary that its final form be static. A well-known illustration of this is the face that a flame is as recognizable a form as a crystal, even though it changes its shape constantly (Atlan 1979). It is the (cyclic) persistence of patterns, rather than fixity of an object in time and space, that is important. This is what one has to keep in mind when considering the choice between the perspective of anomaly and the perspective of anomaly, especially since languages are certainly not fixed objects: they change over time (in technical terms, they are "diachronic"), and this is compatible with both of these otherwise opposing points of view (see Longobardi 2003, Roberts 2007, and the references cited there for diachronic syntax).

These two poles, the one based on analogy and the one based on anomaly, represent the two more radical contrasting views about the source of order in language, and they have been revived at many points throughout the history of linguistics. The idea that either force is what triggers order has been extended to other domains of inquiry and this opposition can be utilized to formulate radical ways of explaining how the parts of the world became structured, and perhaps even how the entire universe became a cosmos (in the original sense of the term: a well-ordered whole). Astrophysics certainly takes this approach, but other disciplines do as well. Biology, for example, is a particularly rich domain for testing the analogy thesis. One of the most interesting viewpoints on this issue is the one adopted in an infrequently cited paper by Alan Turing (1952). Turing was trying to capture the chemical basis of morphogenesis—that is, the biological instructions that may determine the anatomical structure of a resulting organism: "It is suggested that a system [...] although it may originally be quite homogeneous, may later develop a pattern or structure due to an instability of the homogeneous equilibrium, which is triggered off by random disturbances. [...] The investigation is chiefly concerned with the onset of instability. It is found that there are six essentially different forms which this may take" (Turing 1952, 5). Turing realized that his system may account for the tentacle patterns on the hydrozoan polyps known as hydras, which involve whorled leaves, gastrulation, phyllotaxis, and dappling: all specific patterns resulting from the interaction of a few simple instructions in a highly complex system. He also suggested that stationary waves in two dimensions could account for the spectacular phenomenon of phyllotaxis. This is a very surprising theory, especially since Turing's paper appeared a year before the discovery of the structure of DNA by Watson and Crick (1953)—although, interestingly, after Erwin Schrödinger's 1944 essay, *What Is Life?*—and it was one that could tentatively be regarded as espousing a neo-anomalist point of view, since the microscopic perturbations in another wise two-dimensional flat land are what generate the variety of three-dimensional structures observed.

However, besides these philological speculations from the Hellenistic and Roman periods (see Moro 2016 and the references cited there), there are also interesting independent cases in which the notion of analogy or anomaly has played a role in linguistic studies. The notion of analogy, for example, has often been invoked in studies of language acquisition (see the criticism in Yang 2011). Children demonstrate the ability to proceed by analogies when it comes to morphology, especially with the inflection of verbs and the declension of nouns (see Halle and Marantz 1996, Pinker 2011, and again Yang 2011). So, for example, when learning a language with irregular verbs a child will often start by producing regular forms on the sole basis of analogy, the way an English-speaking child might start by constructing the word *goed* instead of *went*. The process of analogy, productive as it may be in word formation, falls short when it comes to syntax, a much too complex system to allow for such superficial inferences. The following sentences offer some clear examples: (a) *John is too lazy to read books*, (b) *John is too lazy to read*, (c) *John is too stubborn to talk to Bill*, and (d) *John is too stubborn to talk to*. There is one apparent analogy here: (a) is identical to (b) save for the appearance of the noun (books) at the end of the sequence; (c) is identical to (d) save for the appearance of a proper name (Bill) at the end the sequence: (b) and (d) are just versions of (a) and (c) that are shorter by one word. The sentence in (b), however, is computed as if the object *books* was present; the sentence in (d) instead is completed as if *John*—and not the missing *Bill*—was present and the subject being "the people," a rather surprising fact, since *John* is rather realized as the subject. No analogy holds.

Another interesting potential field is that offered by copular sentences with predicative noun phrases: sentences like *this picture was the cause* which can be summed up more abstractly as *noun phrase – copula – noun phrase*. In these cases, the *copula* is the medieval term used to indicate the verb *to be* and its equivalents across languages (See the appendix of Moro 1997 for a short history of the copula or Moro 2010 for a much more detailed historical account; for the structure of those sentences, instead see Moro 1997 and 2006). These constructions are remarkably challenging across languages, because this very same sequence *noun phrase – copula – noun phrase* can be associated with two sets of completely different properties. If analogy were the real guide to acquisition in syntax, constructing such sentences would be very difficult for a child to do when acquiring a language. One often-cited contrast can be found in the following set: (a) *This picture of the wall is the cause of the riot*; (b) *The cause of the riot is this picture of the wall*. The two sentences appear to be similar; indeed, the sequence of lexical items offers the same two permuting noun phrases separated by the same verb. But as soon as we test locality, we realize how different they are: (c) *Which riot do you think this picture of the wall is the cause of?* versus (d) *Which wall do you think the cause of the riot was a picture of?* Any superficial analogy would fail to capture such a deep contrast. For any given pair of copular sentences in which two noun phrases can be permuted (as in the example above), one of them will not allow extraction from the noun phrase following the copula even if the same noun phrase would allow extraction if it were following a transitive verb. For example, if we replace the copula *is* in our example with a transitive verb such as *to reveal*, (d) could then be rendered as *Which wall do you think the cause of the riot revealed a picture of?* Why would the copula selectively block one and only one case?

The key to resolving this challenging and puzzling theoretical situation was to abandon the central dogma of clause-structure theory: the dogma stating that the sequence *noun phrase—verb phrase* coincides with the basic logical defining schema of all natural sentences, namely *subject—predicate*, where the subject stands for a substance and the predicate for the property to be attributed to the substance, often expressing the tense and other formal features of the predicate (see Moro 1997, 2010, and Moro 2013 for the translation of Moro 1988 originally proposing the unified analysis of copular sentences). A simple example of this sequence would be one like *this picture caused the riot* where the subject is *this picture* and the predicate is *caused the riot* including reference to the past. The reason to abandon this rigid association between categories and grammatical functions comes from copular sentences. In sentences of the type *noun phrase – copula – noun phrase* the distinction between subject and predicate is not reflected in morphology, because unlike any other case, these two functions are not realized by two distinct categories of noun phrases and verb phrases, respectively: the predicate can be realized by the same morphological category as the subject, i.e., a noun phrase. Thus, in *This picture was the cause of the riot, this picture* is the subject, *the cause of the riot* is the predicate and *was* expresses reference to the past. Clearly, *this picture* and *the cause of the riot* belong to the same lexical category: they are both noun phrases.

This striking fact allows us to abandon the dogma concerning the structure of sentences and to dissociate the sequence *noun phrase – verb phrase* from the ubiquitous and traditional interpretative schema fixing the two fundamental grammatical functions in the rigid *subject – predicate* sequence. More precisely, we can understand the sequence *noun phrase – copula – noun phrase* as capable of realizing either the sequence *subject – copula – predicate* (canonical sentences) or the sequence *predicate – copula – subject* (inverse sentences). In inverse sentences, the post-verbal noun phrase does behave like a subject; for example, extraction is equally banned in both cases: *Which wall do you think a picture of scared me?* and *Which wall do you think the cause of the riot was a picture of?* Again, another hidden analogy gets uncovered as in the case of the impossibility to extract from adjuncts as in the one discussed in a previous chapter: *Who does John want to hire Mary before meeting* and *Of which city does Mary want to describe a planet on a bridge?*

The anomaly of copular sentences justifies a radical revision of the theory of clause structure derived from the unified theory that captures these recalcitrant pairs. It is worth noting that abandoning the rigid interpretation of clause structure does not imply that we should abandon the *subject – predicate* schema, which is still considered to be at the very core of human syntax and a valid prerequisite to possible languages; nor does it imply that we should abandon the asymmetric architecture of clauses which still includes noun phrase verb phrase sequences. The only point to be made here, and it is a far reaching one, is that the *subject – predicate* schema must be dissociated from the *noun phrase – verb phrase* sequence although they must both be separately maintained. Or, in other words, there is no unambiguous mapping between the two.

As for the role of analogy in building linguistic structure, there has yet to be any evidence of a child who has made the kind of mistake to be found in a phrase like *Which wall do you think the cause of the riot was a picture of?*—a mistake, that is, which violates the unified theory of copular sentences. The situation is very different in case studies of the generation of irregular verbs based on the analogy strategy: analogy appears to be a much more powerful, even if marginally misleading, construction system for morphology than for syntax. It is easier for a child to construct *scared* out of the verb *to scare* (for the same reason that it also means a greater likelihood of that child using that same morphological construction system to construct *goed* from *to go*) than it is for that child to apply similar analogy strategies in constructing syntax—because syntax analogy is quite misleading. On the other hand, the notion of anomaly, much in the way Turing understood perturbation to function within a complex system, turned out to be very productive in understanding how we construct syntax. This is, in fact, how we can understand the origin of the present model for syntax, derived from Chomsky (1981): a system that utilizes a recursive combinatorial rule, in which very small differences among lexical features interact with each other generating complex structures. This view has recently been refined to propose that the system also includes symmetry breaking phenomena along with interaction with morphological features (see Moro 2000, 2009, 2013, Chomsky 2013, Rizzi 2015 and the references cited there).

The search for conditions that would restrict the class of possible languages has not been a promising one outside the language domain. Exploring similar cognitive domains is a very interesting line of inquiry, but more to emphasize the differences to be found than the similarities. Moreover, it leads us to find ourselves in a situation typical in empirical science, in which something is easy to describe but very hard to explain. But at least we can now safely exclude what cannot be the source of the regularities constituting the boundaries of Babel: crucially, linear order and analogy—a very unexpected and sharply counterintuitive result.

8

THE SOUND OF THOUGHT

Why do we include the sounds of words in our thoughts when we think without speaking? Are they just an illusion induced by our memory of overt speech? Questions like these have long pointed to a mystery: a mystery relevant to our endeavor to identify impossible languages, but also relevant from a methodological perspective, since to address it requires our radically changing our approach to the relationship between language and the brain. We need to shift from identifying *where* neurons are firing to identifying *what* neurons are firing—that is, from identifying (by means of neuroimaging techniques) where language processing takes place in the brain to deciphering what code neurons exploit when these mental computations take place when we engage in linguistic tasks. This is not to say that the *where* question is trivial or secondary with respect to the *what* question, nor is it to critique the effort of locating language structure in the brain (as was partially suggested in Embick and Poeppel 2005; see also Moro 2015 and Poepple 2014 for general remarks on this matter). Such would be the case if the sole aim of this research perspective were to associate the activation of an area or a neuronal network in the brain to some (cognitive) activity (a truism that has a whiff of phrenology to it). The *where* problem and its correlated technology is not at all trivial if it can be used to provide independent evidence to help choose between two (or more) competing linguistic hypotheses: to show, for example, that non-recursive rules do not activate the language networks in the same way recursive languages do. With that understood, and capitalizing on the results of approach we took to the *where* problem, we can now approach the *what* problem.

Let's begin with a simple question: What is language made of? The immediate answer usually refers to language structure: language consists of words and rules of combination. While this is certainly a good answer it may not be the unique one, for the question can also refer to the concrete stuff of which language consists: that is, what is language made of from the point of view of physics? Language exists in two different physical spaces: outside our brain and inside it. When it lives outside our brain, language consists of mechanical waves of compressed and rarefied molecules of air (i.e., sound); when it exists inside our brain, language consists of electric waves that are the channel of communication neurons exploit for this type of task. Waves: in either case, this is what language is physically made of. The question then naturally arises as to what the relationship is between these two such different families of waves. A priori, we know that there has to be a relationship, since the sound produced when uttering sentences is programmed by neurons controlling the articulatory apparatus and neurons also decipher the content of a sentences via the auditory apparatus; but there is no a priori information on the nature of this relationship or on how and if one can discern a structure common to the two types of waves. Any answer to the question of how similar these two types of waves are would of course shed light on the distinction between possible and impossible languages, and it would do so from a new perspective: one that comes from understanding the *what* problem.

There is one obvious connection between sound waves and the brain. Sound is what allows the contents of one brain, as expressed in words, to enter another brain. Strictly speaking, this is not completely true, because there are other ways for two brains to exchange linguistic information: through the eyes, which is the method you are utilizing as you read this, or through organized and grammaticalized gestures such as to be found in sign language, or through tactile sensations as in Braille, or by touching the face of the speaker with a hand as in the "Tadoma method" used with persons who are unable to see or hear from birth. The situation gets complicated: in the case of reading an alphabetical writing system (as opposed to an ideographic system), for example, the memory of sounds associated with lexical items is encoded in electromagnetic light waves in the shape of alphabetical letters: writing and reading is what this process is about (along, again, with ideographic writing, which does not immediately refer to sound). Although writing is an extremely interesting issue, we can put it aside for a moment and concentrate on spoken language, since it isn't necessary for language to include writing, which developmentally comes later in any case, be it phylogenetically (i.e., in the development of our species) or ontogenetically (i.e., in the development of us as individuals; see Dehaene 2010 and Magrassi 2010). Sound enters us through our ears, traveling across the tympanic membrane, the three ossicles, and the Corti organ in the cochlea, a snail-shaped organ that plays a crucial role in this process. This complex system translates mechanical vibrations of the acoustic signal into electric impulses in a very sophisticated way, decomposing the complex sound waves into the basic frequencies that characterize them (see Rosen 1992). The different frequencies are then mapped onto dedicated slots in the primary auditory cortex. At this point the sound waves are replaced by electric waves. A major discovery was that the electric waves of the acoustic areas correlate with sound waves; more explicitly, the shape of a sound wave is partially preserved in the shape of the electric waves of the neurons in the acoustic areas (see Pulvermüller et al. 2006; Nourski et al 2009, Giraud and Poeppel 2012; Pasley et al. 2012, Bouchard et al. 2013, and Mesgarani et al. 2014). This is not surprising, given that, since the pioneering works by Lord Adrian (see Adrian 1947 for a comprehensive treatise), if not earlier, we have known that no physical signal is ever completely lost when it reaches the neuronal networks; this is what happens to visual signals and there is no reason why it should not happen with sound signals as well. But it is by no means trivial to empirically detect it, because the sound waves could have changed shape when they were translated by the neurons and would thus be unrecognizable through quantitative inspection. So it is quite surprising to discover that electric waves do preserve the shape of their corresponding sound waves in non-acoustic areas, such as, for example, in Broca's area (see Kubanek et al. 2013).

These findings shed important light on the relationship between sound waves and electric waves in the brain, but almost all of them rely on one aspect of the neuropsychological processes related to language; namely, sound emission decoding. This is just one partial view, though, since, as we noted, we know that language can also be present in the absence of sound, when we read (as what we are most probably experiencing at this very moment) or when we use words while thinking—in technical terms, when we engage in endophasic activity. This simple fact immediately raises the following crucial question: what happens to the electric waves in our brain when we generate a linguistic expression without emitting any sound? This was the core question we addressed in the experiment I will concentrate on here (Magrassi et al. 2015a; see also Moro 2015 for a nontechnical presentation). The essential part of the endeavor was to compare the shape of the electric waves characterizing the activity in Broca's area with the shape of the sound waves—not just when speakers were hearing sound, but also when they were reading linguistic expressions in absolute silence; that is, when the input was not acoustic at all. The results were unexpected, to say the least. Analyzing inner speech is by no means a novelty in neuropsychology, as we know from sources ranging from Lev Vygotsky's speculations on psychological development (see the collection in Vygotsky 1986) to analyses based on neuroimaging (see McGuire et al. 1996, among others), to say nothing of medieval reflections like those of Saint Augustine in his *Confessions*. But I would like to illustrate a new means of exploring this phenomenon. First, however, let me briefly illustrate the technique adopted.

In this experiment, data were collected by means of so-called awake surgery (see Calvin and Ojemann 1994 for a full, nontechnical description of this technique and a case study). This technique offers the possibility of stimulating and analyzing the electrophysiological cortical activity of patients who have been awakened after a portion of their skullcap was removed. The invasive nature of this technique, the fragility of the organ involved, and the cooperation of patients who are in an extremely delicate emotional state make this type of research very difficult for obvious psychological, technical, and ethical reasons. Clearly, this procedure must above all benefit the patient, but the benefits are clear. The surgeon who cuts the cerebral cortex to remove a tumor, for example, cannot know in advance (except in specific cases) whether cutting the cerebral tissue will interrupt a neuronal network and thus impair or destroy a cognitive, motor, or perceptual capacity that is supported or conveyed by that network. To minimize any potential damage from the surgery, then, once the patient has been anesthetized and a portion of the skullcap has been removed in order to access the surgical site, the surgeon wakes the patient (for a short transitional period, about ten to twenty minutes) and asks him or her to perform some simple tasks that should require their utilizing the exposed cortex. As they perform them, the surgeon stimulates the patient's cortex by means of small electrodes (which causes no pain since there are no pain receptors in the brain). If the electrical stimulation in a certain portion of the cortex interferes with the performance of a given task, the surgeon knows that cutting that fragment of cortex could permanently damage the patient and can evaluate whether an alternative surgical site is available. The patient thus gains an invaluable advantage from these exercises, and one that is at present practically impossible to obtain through any other technique. At the same time, this technique provides us with a unique opportunity to investigate brain functioning and obtain extremely important data. First, the surgeon can establish the position where a crucial node of a neuronal network associated with a specific task is located in any given patient, which neutralizes one of the major problems related to neuroimaging techniques: the fact that subjects may vary considerably as to precisely where a certain function is carried out in the brain. Moreover, the surgeon may obtain unique data on cerebral neurophysiology, by recording with progressive precision neuronal electrical activity down to the level of a single neuron, although this level is only reached in extremely rare cases with current technology. Increasingly, this technique has been used for pathologies other than focal lesions—for example, cases of pharmacologically intractable epilepsy. In such cases, the surgeon can also implant temporary electrodes that, once the skullcap has been closed, provide continuous information for a lengthy period of time in an everyday environment, and information that is not limited to the scope of the operating room. This measuring method offers us a further step forward in comprehending the neurophysiological processes taking place in the brain: it provides a more precise and defined level of spatial resolution than what neuroimaging techniques are capable of, it avoids the indeterminacy of statistical localization, and it provides specific measures of electrical activity not available through indirect metabolic or perfusional means of measurement.

Let us now turn to our experiment. Sixteen patients—an anomalous number, since this type of experiment is normally run with one or two patients—were involved, but the larger number was justified by the need for a robust empirical basis to confirm and support the quite surprising data we were obtaining. The patients read linguistic expressions aloud: either isolated words or full sentences. We compared the shape of the acoustic waves with the shape of the electric waves in a non-acoustic area such as Broca's area, and we were able to observe a correlation (which was not unexpected). The second step was crucial: we asked the patients to read the linguistic expressions without emitting any sound; they just read them in their mind. By analogy, we compared the shape of the acoustic wave with the shape of the electric wave in the same non-acoustic area as before (Broca's area). It is extremely important to note that a signal was indeed entering the brain, but it was not a sound signal; instead, it was the light signal carried by electromagnetic waves—or, to put it more simply, a signal conveyed by the alphabetical letters we use to represent words (i.e., writing): definitely not an acoustic wave. As for the individuation of Broca's area, we employed a functional strategy—that is, we individuated this area by stimulating the patients' cortex with a light electric impulse. The point at which speech arrest was induced was by definition considered Broca's area. It is worth noting that awake surgery has enabled researchers to realize that Broca's area varies from individual to individual, moving even up to 8 cm around an average center (see Ojemann, Ojemann, Lettich, and Berger 1989; Keller et al. 2009). From this point of view, the advantage of this technique over functional neuroimaging, particularly in terms of spatial recognition, is dramatically evident.

All in all, this was the core result obtained in the experiment: *unexpected as it is,* the shape of the electric waves recorded in a non-acoustic area such as Broca's area when linguistic expressions are being read silently preserves the same structure as those of the mechanical sound waves of air that would have been produced if those words had actually been uttered. The two families of waves where language lives physically are then closely related—so closely in fact that the two overlap independently of the presence of sound. The acoustic information is not implanted later, when a person needs to communicate with someone else; it is part of the code from the beginning, or at least before the production of sound takes place. It also excludes that the sensation of exploiting sound representation while reading or thinking with words is just an illusory artifact based on a remembrance of the overt speech.

One important point must be mentioned here: it is clearly impossible for sound representation alone to contain all the pieces of information necessary to process a linguistic structure. All human languages, for example, contain nonlocal dependencies that are not signaled phonologically in the words filling the space between any two elements creating the dependency. Consider, for example, a simple case like *Mary whom these boys love is very smart*. There is a dependency between *Mary* and *is* and it is nonlocal because the two elements are separated from each other): they both agree in number, among other things, because they are both singular. This information must reach the verb *is* in a way that is separated from the sound of the words between *Mary* and *is*; in fact, if anything, the words preceding *is* are plural (*these boys love*). And, of course, besides nonlocal dependencies, the fact that sound cannot exhaust the linguistic information is also supported by the existence of ambiguous sentences. When one says *I saw Galileo with a telescope*, the sentence has two structures, hence two interpretations: one where *with a telescope* is part of the object of the verb *see* (the individual named *Galileo* was carrying this tool), and the other where *with a telescope* pertains to the meaning of the verb and specifies a certain modality of seeing (by means of a telescope I was able to see the individual named *Galileo*). So to say that the electrical activity of neurons elaborating linguistic information in non-acoustic areas preserves the shape of the sound waves by no means implies that linguistic information is reducible to sound representation solely; that would be like denying the very existence of syntax and reducing language to pure air vibrations.

So where is the syntactic and morphosyntactic information constituting a linguistic expression processed? Prima facie, there are two logical possibilities.

One possibility is that this type of information is processed by some other non-acoustic network in the brain, not by the traditional high-level network that includes Broca's area. In practical terms, this means that the electrical activity recorded in our experiment only pertains to sound (i.e., speech) not to structure (i.e., language). This would be quite a problematic situation since it would conflict with the results of previous experiments showing that the network that includes Broca's area is indeed sensitive to the distinction between a recursive versus non-recursive syntax, which has nothing to do with sound (at least, not inherently). The alternative possibility is that Broca's area is indeed included in a network which computes syntax and morphosyntax *and* that this type of information is meshed together with the acoustic information. Interestingly, the recordings obtained in Magrassi et al 2015a also contained another crucial piece of information compatible with this latter possibility. The number of electrodes showing a significant correlation between sound waves and the electric activity surrounding Broca's area depended on what type of linguistic stimulus was being applied: it was higher if the sound was associated with single (long) *words* in isolation, and it was lower if the sound was associated with (short) *sentences*; clearly the presence of a syntactic structure makes the difference. This has an immediate consequence: it proves that the electric activity we measured in that area was not simply related to speech. It had to be language proper, because words and sentences are not assigned any special sound marker. In a language like Italian, for example, the acoustic expression *telefono* can either mean the electronic tool used to communicate (the telephone proper), or mean "I am making a phone call"—that is, it can be either a simple word (a noun) or a full-fledged sentence, involving a silent first person subject. But it also has a second consequence, and a potentially far-reaching one. Those electrodes located in the surroundings of Broca's area which do not show a significant correlation between sound representation and the electric activity may provide us with information on the syntactic and morphosyntactic properties of the linguistic structures being utilized and ultimately identify the electrical activity corresponding to the recursive structure of syntax. As for the idea that the correspondence between sound waves and electric waves is broader than expected in non-acoustic areas, even when no sound is produced, a further experiment has been carried out which showed that sound-correlated activity during language generation is present in multiple areas of the frontal lobe, which reinforces the impression that our current models of interpretation may not be sufficient (see Magrassi et al. 2015b).

To sum up, the discovery that these two independent families of waves of which language is physically made strictly correlate with each other—even in non-acoustic areas and whether or not the linguistic structures are actually uttered or remain within the mind of an individual—indicates that sound plays a much more central role in language processing than was previously thought. From a methodological point of view, the paradigm adopted here constitutes a genuine example of our shift from the *where* to the *what* problem. It is as if this unexpected correlation provided us with the missing piece of a "Rosetta stone" in which two known codes—the sound waves and the electric waves generated by sound—could be exploited to decipher a third one: the electric code generated in the absence of sound, which in turn could hopefully lead to the discovery of the fingerprint of human language (i.e., the discrete infinity provided by syntax). It goes without saying that this discovery raises a lot of new questions, but on the other hand, this is always a good sign in science, for it is the number of new questions raised by a theory rather than the number of ultimate answers to old questions that manifests the strength of a research program. Some of the new questions include the following: What kind of electrical activity is elaborated in a language network (one that includes Broca's area) by persons who have never been able to hear any sound from birth? Can we exploit electro-cortical information to access the linguistic thinking of aphasic patients whose articulatory apparatus alone has been damaged, and hear them speak again, albeit through an artificial device? Can we get a better understanding of language used in dreaming or in patients who are in a minimally conscious state? (For language and dreaming, see Moro 2014 and references cited there.) Can we consider severe stuttering as a form of miscoordination between different sound representations in different networks and hope to intervene and cure it? Can these discoveries lead to an unethical use of devices to excerpt linguistic thought from people who do not want to communicate it?

There are also interesting theoretical ramifications, and these are strictly connected to the individuation of impossible languages. In fact, this discovery unexpectedly suggests potential research paths that may lead to a unification of the physical properties of language with its formal ones. Ideally, the two major properties characterizing syntax—recursion and locality—should not only be compatible with this code, but also be derivable from it. The presence of soundlike waves in the core non-acoustic areas of language may thus affect the structure of possible languages and offer new empirical data that might allow us to understand the properties of syntax as being the optimal conditions needed for computations of this type of wave to be made. This is perhaps the first time that a physical property has opened itself up to exploration toward understanding the formal structure of a human language.

The very fact that the majority of human communication takes place via waves may not be a casual fact; after all, waves constitute the purest system of communication since they transfer information from one entity to the other without changing the structure or the composition of the two entities. They travel through us and leave us intact, but they allow us to interpret the message borne by their momentary vibrations, provided that we have the key to decode it. It is not at all accidental that the term *information* is derived from the Latin root *forma* (shape): to inform is to share a shape.

In his *Philosophical Investigations*, Ludwig Wittgenstein (1953, 344) once asked: "Is it conceivable that people should never speak an audible language, but should nevertheless talk to themselves inwardly, in the imagination?" The results of this experiment unexpectedly revive this prophetic question under a new light, and more importantly, they suggest new questions altogether.

9

THE INVERSE THUNDERSTORM

Being out in the countryside during an intense thunderstorm is quite an experience. Lightning flashes around us with an unpredictable rhythm while thunder roars from a distance. We are immersed in waves of light and air that arouse intense sensations. Imagine, though, an experiment that inverts the paths of the optical and auditory nerves: "With the nerves of vision and of hearing severed, and then crossed with each other, we should with the eye hear the lightning-flash as a thunder-clap, and with the ear we should see the thunder as a series of luminous impressions." Strange as this idea may be, it is also a natural consequence of the new perspective our era has given us on our cognitive capacities. Such a perspective must have a counterpart in our language capacities; since we've identified some neurobiological pillars for the boundaries of Babel, we can conceive of a neurobiological structure that allows a human brain to acquire, manifest, and use language in the same way a neurobiological structure allows for other capacities, such as the sensory capacities to see and hear. It is quite reasonable to imagine a possible rewiring of language with other capabilities. What is striking is that the sentence I quoted above was written by Emil du Bois-Reymond in 1874, only thirteen years after Broca published his

famous article. The line of inquiry Du Bois-Reymond was opening up, though, has been practically ignored, at least from an
experimental point of view, and particularly so in the case of
language. Lenneberg's caveat explains why: because languages
had been considered to be arbitrary, cultural conventions, they
could not be treated on a par with other brain-driven phenomena such as vision, for example. If thunder had been considered arbitrary, Du Bois-Reymond would never have written that
sentence. But years prior to today's accumulation of discoveries
on the relationship between language and the brain, another
opinion was expressed that constituted an ideal complement to
Du Bois-Reymond's idea: "The information provided by lexical
items and other expressions yields perspectives for thinking and
speaking about the world by virtue of the way their elements are
interpreted 'at the interface'; embedded in different performance
systems in some hypothetical (perhaps biologically impossible)
organism, they could serve for some other activity, say, locomotion" (Chomsky 1993). This extreme view, which is now sort
of a null hypothesis as to the biological nature of language, did
not emerge from neurobiological studies, but was deduced from
comparative linguistic data—derived from the isolation of minimal primitive elements and the basic rule(s) of recombination—
as well as reflections on language acquisition. What matters for
our purposes is that the notion of a possible language is somehow linked with the notion of a possible organism; in fact, the
implicit assumption is that language can arise only as the result
of a possible organism, in the genuine biological sense.

What seems totally missing from this point of view and its
contemporary revival is the fact that we do not know what leads
syntactic rules to take on the form they have. Why, for example, aren't they based on the only uncontroversial fact about

language—the fact that words come one after another in a linear sequence? Putting it more simply, why do we consider it natural that in the sentence *It's in the garden that the girls who sing eat*, the garden is the place where the eating takes place rather than the singing, even though the word *garden* is closer to the verb *sing* than it is to *eat*? To date, we have lacked the empirical and theoretical arguments to explain this, even though we can describe it in terms of (recursive) hierarchical rules and exclude the notion that the linear sequence of a sentence is what matters to establish actual syntactic rules (i.e., dependencies and locality restrictions). To describe something like this, though, is not to explain it, and we cannot explain the preference of hierarchy over linearity. In a sense, we are like archeologists of the future who happen to find a few electronic keyboards. Such archeologists might be curious as to what led to the QWERTY layout, since alphabetical order is a better-known, more common way of arranging letters. This layout was chosen by the designers of mechanical typewriters as one of several potential solutions to keep the metal typebars from clashing when the typist struck the keys too quickly; some letters that are alphabetically close to each other are often used together and this would cause a typebar returning to its original position to clash with another that was moving forward to press against the inked ribbon. If archeologists of the future have no knowledge of or access to a mechanical typewriter, though, it would be completely impossible for them to decode the mystery of the QWERTY layout, which would thereafter pose an enigma. A similar scenario may affect the search for the origin of language.

Let us rephrase the core question: why do the syntactic rules of human language ignore linear order and capitalize on a hierarchical architecture based on the recursive application of a binary combinatorial rule? This question now appears similar to the one faced by future archeologists staring at the QWERTY keyboard. We not only lack reasons but, probably, also the possibility of solving this riddle, be it at the level of phylogenetic or ontogenetic development—that is, be it a question of the story of our species or the development of our organism as individuals. In a sense, this scenario is to be expected in modern evolutionary theory. Consider the notion of *exaptation*, a term proposed by the biologist Stephen Jay Gould (Gould and Vrba 1982; Gould 1997). This term describes a scenario in which a certain characteristic of the organism (technically, a phenotypic feature), selected by evolutionary pressure related to a certain function, is successively exploited for a different function and then put under the selective pressure of that different function. In Darwin's words: "When this or that part has been spoken of as adapted for some special purpose, it must not be supposed that it was originally always formed for this sole purpose. The regular course of events seems to be, that a part which originally served for one purpose, becomes adapted by slow changes for widely different purposes" (Darwin 1862, 202). One famous example is that of insect wings. When insect wings first appeared, they were not big enough to support the body's weight and allow for flight. Why, then, did wing surfaces evolve to the point where wings turned into an effective tool for locomotion? Obviously, we cannot claim that random, local mutations could have occurred in anticipation of this future function: natural selection operates in the here and now, it doesn't look ahead. In fact, wings developed as heat exchangers; their vibrations were used to fan the surface of the insect that was normally exposed to high temperatures from the sunlight, and only when they attained a sufficiently wide surface *by chance* through successive casual mutations could their vibrations also be used for flight (see Wesson 1991, among others). Two different evolutionary steps involving the same biological structure, one following the other, were governed by a response to two different functions that only marginally overlapped for accidental reasons: cooling and movement. In a sense, the old Lamarckian adage according to which "the function creates the organ" is turned upside down: rather, a function develops (or a need is expressed) when the organ is significantly modified by chance for other reasons.

The case of the QWERTY keyboard is, so to speak, the inverse of exaptation. Call it "kataptation" for analogy with exaptation (see Moro 2013): there is a feature, selected for reasons unknown, which does resist disappearing, even if it is no longer useful, even if the function it was selected for has been abandoned or has become completely irrelevant and no new function is born or manifested by that feature. This feature just ended up being tolerated, functioning in equilibrium with, and interacting with, other features. There could be any number of reasons for this. In the case of the keyboard, the persistence of the QWERTY layout was due to the industrial cost of changing a model and the habituation of typewriter users who had undergone special training involving teachers, books, schools, and so on. Industries would not have been willing to propose a new keyboard layout when such a change could render their product unprofitable. If we adopt this point of view in biology, we can regard the structure of human languages merely as a case of kataptation. This could either be due to the way the brain is wired or to the nature of the signal; certainly, it cannot be traced back to communicative factors, for other systems could be designed in a much better way. Interestingly, the idea that this structure could in principle be related to the physical nature of the signal may lead to some novel speculation, given the discovery that the code exploited by neurons in non-acoustic areas maintains the shape of their corresponding sound waves. Sound waves impose some restrictions on syntactic rules that prioritize hierarchical organization over sequential organization, because of the way the waves match each other and condition the processing of syntax. It is of course too early to approach these issues from this perspective, but the possibility has been raised and has to be considered.

Another potentially interesting way of approaching impossible languages from a biological point of view is, of course, via genetics. Before touching on this field, though, let us consider the following statement by the biologist Peter Medawar (Medawar and Medawar 1983, 9):

> One of the gravest and most widespread aberrations of geneticism is embodied in the belief that if any characteristic is enjoyed by all individuals of the community, it must be genetically underwritten. Thus, if it should turn out that a certain basic linguistic form such as the Aristotelian subject/predicate form is an element of all languages of the world, then its usage must be genetically programmed. (Some of Noam Chomsky's writings are not guiltless of this assumption, which is also a disfigurement of sociobiology as it steers its precarious course between the twin perils of geneticism and historicism.) It may be well to repeat in this context the reason why the supreme canon of geneticism is not satisfactory: if any trait is to be judged "inborn" or genetically programmed, then there must be some people who lack it. The ability to taste phenylthiocarbamide, for instance, is known to be genetically programmed because there are those who lack it.

If we take this caveat seriously, we can only conclude that we are still a long way away from having a genetic approach to language structure, and I do not mean molecular genetics: even at the very basic level of phenotypes no one has ever found anyone to be lacking some of the basic features of syntax, the proto-typical basic binary combinatorial recursive operation. We not only lack a genetics approach to language structure, we even lack a Mendelian one, that is one where a linguistic equivalent of phenotype—in first approximation, the observable trait of living organism such as the color of the eyes in humans—is expressed, suppressed or graded across generations with the same statistics governing phenotypes in living organisms (Miko 2008 and Watson et al. 2014). There could be many explanations for this. One is that if an individual lacks the capacity for syntax, other cognitive faculties are activated to overcome the impairment, which is what happens with dyslexia in certain cases (see Brambati et al. 2006 and Swagerman et al. 2015). Or there could be another much more radical reason. It could be that the gene pools expressing language express themselves in organs indispensible for life; the lack of one of these genes may then just result in the premature death of the fetus. This would provide the human species with a powerful albeit quite indirect defense, so to speak: a defense against any mutants equal to humans in every way but for the capacity to develop a comparable language—that is, a language with the capacity for recombining discrete lexical elements ad infinitum in a recursive way. In other words, it is as if to produce a human language the entire genome required to produce an individual is required: we are our language.

Descartes captured this coincidence in a famous passage in his *Discourse on Method:* "There are no men so dull and stupid, not even idiots, as to be incapable of joining together different words, and thereby constructing a declaration by which to make their thoughts understood; and ... on the other hand, there is no other animal, however perfect or happily circumstanced, which can do the like" (Descartes 1637). Note that by "idiots" Descartes was referring to pathological conditions. For whatever reason, human language is not just the fingerprint of our singularity; it is also so deeply rooted in us, and inherently so, that no human can exist without it.

We have seen how the distinction between possible and impossible languages may bear on the restrictions imposed on living organisms. At this point it is indeed more speculation—a vague intuition, perhaps—than a basis for any concrete experimental paradigm; but however vague and speculative this approach may be, new empirical, neurobiological evidence is beginning to emerge against a theory that traces the birth and development of language in our species back to any general prelinguistic need such as communicative functions or to the emancipation of motor planning. This deepens the sulcus between possible and impossible languages if the former are to be conceived as instruments invented for better communication.

10

BETTER THAN POSSIBLE: ARTIFICIAL LANGUAGES

Being possible doesn't necessarily mean being optimal, and this is obviously the case for languages. For this very reason the dream of creating better languages arose as soon as languages could be considered with sufficient rationality. Moreover, the invention of artificial languages is an ambition that has often intersected with that of those who have sought the *Ursprache*—that is, the original language, the one which came before all others, before the catastrophe of the Tower of Babel; the common idea that time consumes and corrupts things and minds often implied that the original language was also the perfect one (see Eco 1993 and references cited there). Although the temptation to create—or the hope of creating—a better language has in fact never led to satisfactory results, it has always produced an interesting laboratory of ideas, because to desire a better language one must first understand how possible languages have failed. This, of course, depends on what one aims to do with a language, which provides us with an initial natural partition for the taxonomy of possible artificial languages. Essentially, it boils down to at least four empirically and theoretically distinct goals and domains: (1) better communication (languages designed to either contrast or facilitate the information being transferred, in terms of speed, simplicity, or accuracy); (2) experimentation (languages exploited as tools for investigating the structure of natural languages in the human mind/brain); (3) disambiguation (essentially philosophical languages aiming at reducing the ambiguities of natural languages, especially those captured by logical formalizations); and (4) pleasure (languages that can either evoke imaginary worlds or simply provide a powerful descriptive feature capable of wringing the heartstrings of cultural identity). Each of these goals can be subdivided in turn into different subtypes. For example, experimentation may include either testing—as with experiments exploiting non-recursive syntactic rules—or simulation, as in the case of languages designed to allow electronic devices to imitate language understanding and production and interface with human subjects. These considerations also lead to another interesting notion, the notion of a "probable language." This is a language that would be better suited for certain external conditions: for example, articulatory, physical, environmental, or social conditions (see Newmeyer 2005, among others).

The land of invented languages covers a vast portion of the domain of Babel; it is such broad terrain that many books would be necessary to adequately cover the topic. Nevertheless, two major aspects of this domain are relevant to the distinction between possible and impossible languages and so should be mentioned here: first, the tension between expression and thought, and second, the facilitation of communication. I will approach these two topics with some simple case studies.

Let us start with an example of the tension between expression and thought. As noted earlier, we all know that languages contain ambiguities. A sentence like *They are flying planes* may either mean that the objects I am pointing to are planes and they are now flying, or that some individuals are now piloting some planes. There can be many cases of interesting ambiguities at the logical level as well, as with *John didn't read many books*—which can either mean that John read not many (i.e., few) books or that there are many books that John did not read, even if he is an avid reader. Such ambiguities are pervasive and their existence has provided a powerful impetus to the study of language. One typical case study comes from so-called existential sentences, or *there*-sentences—a specific type of constructions of the type *there be* noun phrase, such as *there is a beautiful proof of this theorem* (see Moro 1997, 2006b and references cited there for a detailed analysis). These sentences challenge the analysis of ambiguities in a very sharp way and have had a pervasive impact on our understanding of the overall design of grammar. Consider for example the following contrast: *there aren't many girls* versus *there aren't pictures of many girls*. In the latter sentence, we face an ambiguity which is similar to the example we just saw; it can either mean that there are many girls such that there aren't pictures of them or that there are pictures of not many (i.e., few) girls; whereas the first sentence can only mean that there are not many (i.e., few) girls: it is not ambiguous. Why this is so is not at all trivial to explain and it shows that ambiguities are sharply affected by that unreasonable sieve we have seen at work in syntax. *There*-sentences are without question problematic.

One of the most striking cases of ambiguity in the entire history of Western linguistics—dominated by Indo-European language models—involves the verb *to be*. The history alone of the name generally given to this verb since the Middle Ages (the *copula*) deserves attention. I have already written extensively on this history and the colossal bibliography surrounding this specific issue (see the appendix of Moro 1997 and Moro 2010);

however, the reason I bring it up here is that a seventeenth-
century philosopher, Juan Caramuel y Lobkowits, Bishop of
Vigevano in Northern Italy, published a treatise in 1681 describing the invention of a new artificial language—the *Leptotatos*,
or *subtilissimus* ("the most subtle," in Ancient Greek and Latin).
This was a metaphysical dialect meant to remedy the deficiencies of existing languages, deficiencies typified by the alleged
different meanings of *to be*. To avoid ambiguity, Caramuel proposed that the five basic meanings he identified in *to be* should
be unambiguously expressed by different artificially designed
verbs, which were then further distinguished in the most subtle
ways and combined with other elements to generate fifteen different verbs. This is a typical case in which philosophy and logic
intervene to reduce problems caused by a language deemed phil-
osophically and logically deficient. But the solution of multiply-
ing roots in order to make the minimal elements of the lexicon
adhere to the metaphysical structure of a word is the opposite of
the tactic one would expect to be taken with artificial languages,
because such languages are normally designed to be simpler
than actual ones, especially from a morphological and lexical
point of view.

A similar approach to Caramuel's was taken by Bertrand Russell with respect to the same verb, *to be*. In a brave attempt to
repair the damage he himself had caused to the Fregean project
of grounding arithmetic in logic, Russell raised his passionate
voice to declare that "it is a disgrace to the human race that it
has chosen the same word *is* for those two such entirely different
ideas [predication and identity]—a disgrace which a symbolic
logical language of course remedies" (Russell 1919).

It is not overly important that this irreducible ambiguity was
the result of an incorrect analysis of linguistic data, as Jespersen

(1924) first noted (see also Moro 1997 for a transformational analysis of the copula that avoids the ambiguity problem). What really matters here is that language was still considered too weak a system to convey meaning in an accurate and satisfactory way; apart from creating ambiguity, it was also thought to produce vagueness, redundancy, tautologies, and so on. It took the revolution of the formal analysis of human languages, which essentially combined Montague grammar and generative grammar in a unified framework, to realize that human languages do not need to be rescued by logic. There is nothing to remedy; it is in fact quite the opposite: human languages are so rich that they can convey meaning with much more complex structures than those generated by logical expressions, and in a compressed fashion (see Graffi 2001 for a historical account of this debate as well as Chierchia and McConnell-Ginet 2001 for an extensive and critical illustration of the fruitful interaction between formal semantics and syntax; on pragmatics, see the influential work of Grice 1975). The "logical form" in the modern sense of the term, then, is no longer a tool to minimize problems generated by human languages, but rather a level of representation in which the inherent richness of linguistic expressions is made explicit and rendered in a more expanded representation.

The second aspect of artificial languages—that is, the idea of facilitating communication—is not completely distinct from the first one. The additional ingredient to consider is the fact that a new, simpler language may be needed among groups of people otherwise not able to understand each other. There are two possibilities here. One is to proceed by choosing the group of languages one wants to help people learn to speak and to create a mixed repertoire of phonemes, roots, inflections, rules of combination, and so on. This is the case with Esperanto and

Volapük, both invented in the last decades of the nineteenth century. These alternatives may be useful for those who already speak the languages the repertoire draws from, but they are certainly of little use to those who speak other languages—the destiny, in Otto Jespersen's words, of a language like Volapük is "sudden success and equally sudden failure" (Jespersen 1928, 33). Another way to facilitate communication—in fact the radically opposite way—is to avoid referring to any language: this is what was behind Leibniz's invention of a *Lingua Characteristica Universalis*, or the contemporaneous and less known *Philosophical Language* of John Wilkins. In a letter to Nicolas Remond in January 1715, Leibniz describes his proposed language as "a sort of *general algebra* in which all truths of reason would be reduced to a kind of calculus. At the same time, this would be a kind of universal language or writing, though infinitely different from all such languages which have thus far been proposed." In this case, obviously, it isn't a matter of finding the minimal elements common among a group of established languages, but instead the minimal elements common to human knowledge. Needless to say, this "philanthropic" project did not meet with success, for this minimal repertoire was and is far beyond our comprehension. Some of the works by the Italian mathematician and logician Giuseppe Peano—namely, the *Latino Sine Flexione* and more importantly, the *Algebra de Grammatica* (see Kennedy 2006 and the references cited there)—represent other excellent examples of a similar effort, involving a search for common roots and a certain degree of linguistic processing in order to ensure better communication. Unfortunately, though Peano's works show an original and surprising intuition in defining the primitive elements themselves in terms of "verbal equations," they do not even come close to capturing the major aspects of human language syntax, including, for example, the locality conditions on dependencies that we have glimpsed in previous chapters.

All of these attempts to design a better language for communication have proposed a reduction of irregular forms, though—which is indeed a rational move and one that challenges the nature of natural languages—as well as a reduction of redundancies. The latter endeavor is a very delicate one. Redundancy in language is ubiquitous, although it can manifest itself differently depending on the specific language. Consider, for example, the following noun phrase in English and its translation in Italian: *This new wonderful red dress* and *Questo nuovo meraviglioso vestito rosso*. In English, the morpheme expressing singular number is overtly marked only once in the word *this* (as opposed to *these*); in Italian, the same morpheme (syncretically expressed with the masculine gender) is repeated five times on each and every word (by means of the vowel /o/). Since there is no reason to suppose that the English linguistic expression is harder to parse or that it allows for more misunderstanding than the Italian one, it is reasonable to think that in creating an artificial language, redundancy could be avoided by looking to English rather than the vacuously redundant Italian. But it is still worth exploring the hypothesis that redundancies may be of some practical use; for example, in minimizing the dispersion of information during transmission, it seems to me that the desire to reduce redundancy may reflect a basic fallacy: what may appear nonsensical or superfluous to an observer may not be so for a computational system. Science contains many examples of such confusion. For many years, for instance, the very fact that certain portions of the DNA molecule did not contain information to code for genes led biologists to think that these segments were not interpretable, and so they dismissed them as "junk DNA" (see Ohno

1972 for the origin of the term). But after further research in the field, it was discovered that these portions of DNA were in fact very useful. For one thing, they can reduce the impact of dangerous mutations by adding inert structures to the molecule, and they provide a sort of punctuation to the reading of DNA by the cellular mechanism of protein synthesis; they do so by creating the right pace for cellular protein production (see Gould 2002 for an in-depth discussion of the expression "junk DNA" and its role in biology). For this very reason, it seems illogical to speak of redundancy, or more generally, of an optimal or perfect amount of structural information in a language, given that we have not decoded the role of elements that may look redundant to an observer. After all, the scope of possible languages has not been fully discovered; we are only approaching the boundaries of variation, and the very existence of redundancy may be one of the prices languages have to pay if they are to be acquired by children.

As noted, another factor should be considered when approaching artificial languages, which springs from the pleasure the human mind experiences in constructing complex and consistent universes made of combinatorial procedures that rely on analogy, symmetry, and variation. Such pleasure can be found in music—Bach's *Goldberg Variations* (BWV 988) are probably among the most interesting examples—or with poetry, architecture, and painting, much in the sense Herman Hesse envisaged in his *The Glass Bead Game*. Language structure and the lexicon are clearly included in this eternal play of cosmic constructions, perhaps even more so than with any other form of expression, since words also have the force to evoke religious, spiritual, and magical values in ways that other symbols may not. After all, the only act of creation God allows humans to perform in the biblical tradition (and which thereby makes humans similar to her or him), is that of giving names to creatures (see Moro 2016 for a comment on this tradition): the act of generating language and language as a whole must have fascinated every culture down through time. The creation of a language has to be regarded as either being the ultimate form of hubris or the most representative and creative aspect of the human mind.

This short detour into the land of artificial and mythical languages ends here. More than nine hundred languages have been invented in Western culture over the last nine hundred years (see Albani and Buonarroti 1994 for a monumental collection of artificial languages up to 1988; also see Okrent 2009). What is certain is that the boundaries discovered for Babel—discrete infinity, recursion, and locality in particular—remain intact even in the face of an artificial Babel, and have proven to be even stronger than the human imagination. Or—for a different point of view—however powerful it may be, the human imagination perhaps needs to be protected from itself; if the imagination is not a completely disruptive force, it may just be because these boundaries offer it a stable grid and a safe containment, which allows creativity to be expressed and communicated at every level.

11

A CLOSER LOOK AT THE TURTLE'S EYES

There is a self-portrait (shown on the facing page) by an Austrian Baroque painter, Johannes Gumpp, who along with his portrait remains practically unknown. He left only two paintings to posterity, each one representing an identical subject, with their only difference being that one is in a round frame and the other is in a rectangular one. The subject of both paintings is a symmetric image: at the center of the painting, we see the artist from behind looking at his face reflected in a mirror on his left while reproducing the same image on his right (*Double Self-Portrait with Mirror and an Easel*, 1646). This may just be a coincidence or the result of my misinterpretation, but this painting seems to me to provide a dramatic representation of our understanding of language. We have two entities, two images of a face, manifested in two different realms: the mirror and the canvas. The two entities have essentially isomorphic structures: the structure of one is essentially reproduced in the other; the major actor, the artist reproducing it, however, remains hidden from our line of sight, so the face at the origin of the two faces visible to us must remain unknown. We can only know the limits of his actions and the possible images deriving from his actions through an indirect perspective and thus through an indirect form of knowledge, not unlike the shadows in Plato's cave. The two faces represent the two domains we can observe: the sound waves and the electric waves, the stuff languages are made of. The more we deepen our knowledge of the structure of language, the more we realize that despite the fact that we approach it with progressive approximation, we cannot reach the fundamental property of language: the creative use we make of it.

This failure should never make us desist from scientific inquiry into language and its biological foundations: the same way that the fact we are unable to grasp the core content of consciousness does not make us desist from seeking to understand the brain and behavior. When it comes to language, we find ourselves in the position of a person who wants to provide a theory of the meaning of the word *caress* but can only manage to define the structure and limits of a movement of a hand: a definition that may also describe a fist but does, at least, exclude a kick. More explicitly, knowing a structure does not imply that we know how the structure is going to be used; however, it is clear that the more one knows about a structure, the more one understands the degrees of freedom it allows for and the more one can circumscribe the range of possibilities that it allows. Excluding creativity from our range of observations on language creates a severe limitation but one that cannot be avoided—for example, it prevents us from evaluating the effect of language on imagination and experience or, to use a very neglected word, on fantasy. *Fantasy* comes from the Ancient Greek verb *phainomai*, which means "to appear in front of us," and language has this power of letting us experience things as if they were in front of us without their actually being so. Dissecting language structure and mapping it into neuronal networks in the brain can sometimes cause us to forget that language is not only an invaluable tool for registering what is real, but it is also capable of reversing the flow and creating a world from the repertoire of independent elements: elements that can form as sublime and visionary an entity as Dante's *Paradiso* or as prosaic and corporal as a menu.

The example of the menu is particularly revealing. Suppose you have a very simple menu that lists espresso, milk, bread, and anchovies: in reading these words, you can easily associate them with the single specific taste you have probably been able to experience when eating or drinking them. You can also combine them: you have probably had an espresso with milk or have drunk a glass of milk while eating bread, and your memories of these actions are activated to provide you with pleasant sensations. But what if I ask you if you would like to have an espresso milkshake with anchovies? You do not need to have experienced it: the combination of tastes in your brain is carried out by the interpretation of sounds—actually, of electromagnetic waves (i.e., written words)—which results in a not-quite-palatable sensation (one would hope). Your brain has just been turned into a virtual kitchen; without actually experiencing it, language has allowed you to evaluate something you have not tried before, and through the simple case of a very rudimentary menu. Using the entire vocabulary of a language allows us to experience a practically infinite range of sensations, surely much broader than a menu list, such as those one might feel when reading, for example, a colossal novel like *War and Peace*. Nevertheless, the creative use of language *qua* creativity lies outside the possibility of a formal description.

Language structure—that is, the discrete infinity of syntax—and the creative use of language are just the two pillars of the singularity of human language. Science has a myriad of puzzles and mysteries; we cannot know in advance if language belongs to one or another of them, but in any case, it is not hard to recognize that language was the big bang for *Homo sapiens*, and that exploring it will tell us important things about ourselves. In one of the most shocking crime stories ever written, a man tries to reconstruct a scene; he says: "Burst out what will, I seek to know my seed, though it be small. ... Being born what I am, I could never be another, so I should fathom all the secret of my descent." This man is the King of Thebes and the lines are taken from Sophocles's most famous—and most perfect, according to Aristotle—tragedy (*Oedipus Rex*, 1076ff.). The reason these words are relevant here is that they single out two conceptually distinct perspectives in the research on any type of origin: the seed and the descent. Studying language is a comprehensive enterprise that provides us with a privileged perspective to understanding the seed and the descent in regards to the human species. I can see no more comprehensive enterprise than this. Whether we will succeed in reaching either goal is not for me to judge. In fact, it seems to me that human language is to us what the turtle was to Achilles: whenever we feel closer to understanding language, our object of inquiry seems to move a little farther off. Nevertheless, the hope is that even if we may not ultimately put our hands on our own turtle, at least we will get close enough to look directly into its eyes.

The journey into the geometric (i.e., formal) architecture of language and the neurobiological activity correlated with it ends here. Have we definitively solved the mystery of impossible languages? Of course not—to my mind, science does not even allow for solutions—but we are now in the privileged position of being able to exclude some potential answers and to refine our questions so as to formulate new ones that were previously unimaginable. There is currently much new speculation on both the geometric and the neurobiological approaches to the matter. I see two major questions rising above the others. The first, from a formal perspective, is whether all principles of grammar could be captured configurationally: that is to say, whether the geometry of language might not be the actual stuff from which language structure is made. Locality as suggested in the so-called "connectedness theory" proposed by Kayne (1984), semantic role assignment as in Hale and Keyser's (2002) approach, movement as in dynamic antisymmetry (Moro 2000), the fine-grained map of clause structure as in the "cartographic project" (Cinque and Rizzi 2008): all converge on the possibility of rendering the entire language structure in a purely geometric way. If this approach were carried out in full, impossible languages would be, so to speak, a sort of "non-Euclidean" grammar—a grammar in which the metric fails for lack of some structural postulate.

The second major question, from a neurobiological perspective, is whether we can exhaustively extract the correlates of morphosyntactic features from the electrical activity of the brain and attain the level of granularity invoked by Poeppel (2014), in conjunction with the advances being made by theoretical linguistics in identifying the minimal primitive elements and operations that constitute language (Chomsky 2013, Rizzi 2015). Ideally, one should get to the point at which a "subtraction" method could apply to the soundlike waves of the electrical activity in superior language areas and make such granularity possible. Should this speculation find some grounding in a successful scientific research project, a unification of the physical and formal properties of language structure could be reasonably entertained.

The theoretical and empirical challenges concerning the exploration of language have completely changed since the idea of impossible languages was adopted as a guideline for research, but the real challenge, in the end, does not pertain to an object: it pertains to us. Language, like theorems and symphonies, exists only in us and it is in ourselves that the "big bang" that matters in the biological history of our species is to be found; outside of us, there are only objects, motion, and light. Constellations and symphonies exist because we exist and we look and listen to them. And so it is for sentences. When we study sentences, we find ourselves in the same situation as someone who studies light. We don't actually see light; we only see its effects on objects. We know it exists because it is partly reflected by the things it encounters, thereby making visible what would otherwise be invisible. It is in this way that nothing, illuminated by another nothing, becomes, for us, something. Words and sentences work the same way: they have no content of their own, but if they encounter someone who listens they become something. We are part of the data.

REFERENCES

The citations in this book have been kept to a minimum. For detailed and updated references in the fields of syntax and neurolinguistics, see Whitaker 1998, Graffi 2000, Ingram 2007, Denes 2011, Bambini 2012, Kandell et al. 2013, and Moro 2013, 2015 and all the references cited in these texts. The notion of "impossible language" in the sense adopted here was inspired by the formulation of grammar first proposed in Chomsky 1962 and refined in Chomsky 1965.

Abutalebi, J., S. Brambati, J. M. Annoni, A. Moro, S. Cappa, and D. Perani. 2007. Auditory perception of language switches: Controlled versus automatic processing as revealed by event-related fMRI. *Journal of Neuroscience* 27 (50): 13762–13769.

Adrian, E. D. 1947. *The Physical Background of Perception.* Oxford: Clarendon Press.

Albani, P., and B. Buonarroti. 1994. *Aga Magéra Difúra: Dizionario delle lingue impossibili.* Bologna: Zanichelli.

Albertini, S., M. Tettamanti, and A. Moro. 2015. Sintassi e working memory: Un nuovo paradigma di valutazione. *Sistemi Intelligenti* 27 (1): 27–44.

Atlan, H. 1979. *Entre le cristal e la fumée: Essai sur l'organisation du vivant.* Paris: Éditions du Seuil.

Baker, M. 2001. *The Atoms of Language.* New York: Basic Books.

Baluška, F., S. Mancuso, and D. Volkmann, eds. 2013. *Communication in Plants: Neuronal Aspects of Plant Life.* Berlin: Springer.

Bambini, V. 2012. Neurolinguistics. In *Handbook of Pragmatics*, ed. J.-O. Östman and J. Verschueren, 1–34. Amsterdam: John Benjamins.

Bar-Hillel, Y. Cybernetics and linguistics. 1971. In *Aspects of Language: Essays and Lectures on Philosophy of Language, Linguistic Philosophy and Methodology of Linguistics*, 289–301. Jerusalem: Magnes Press.

Berwick, R. 1996. The language of the genes. In *Integrative Approaches to Molecular Biology*, ed. J. Collado. Cambridge, MA: MIT Press.

Berwick, R. 2011. Syntax facit saltum redux: Biolinguistics and the leap to syntax. In *The Biolinguistic Enterprise: New Perspectives on the Evolution and Nature of the Human Language Faculty*, ed. A. M. Di Sciullo and C. Boeckx, 65–99. Oxford: Oxford University Press.

Bouchard, K. E., N. Mesgarani, K. Johnson, and E. F. Chang. 2013. Functional organization of human sensorimotor cortex for speech articulation. [Erratum in: *Nature* 498, 526] [2013] *Nature* 495:327–332.

Brambati, S. M., C. Termine, M. Ruffino, M. Danna, G. Lanzi, G. Stella, S. F. Cappa, and D. Perani. 2006. Neuropsychological deficits and neural dysfunction in familial dyslexia. *Brain Research* 1113 (1): 174–185.

Breitbarth, A., and H. C. van Riemsdijk, eds. 2004. *Triggers.* Berlin: Mouton de Gruyter.

Bursill-Hall, G. L., ed. 1972. *Thomas of Erfurt: Grammatica Speculativa. The Classics of Linguistics, 1.* London: Longmans.

Calvin, W., and G. Ojemann. 1994. *Conversation with Neil's Brain: The Neural Nature of Thought and Language.* Reading, MA: Addison-Wesley.

Cappa, S. F. 2001. *Cognitive Neurology: An Introduction.* London: Imperial College Press.

Cappa, S. F. 2012. Imaging semantics and syntax. *NeuroImage* 61 (2): 427–431. doi:10.1016/j.neuroimage.2011.10.006.

Cappa, S. F., A. Moro, D. Perani, and M. Piattelli-Palmarini. 2000. Broca's aphasia, Broca's area and syntax: A complex relationship. *Behavioral and Brain Sciences* 23:27–28.

Caramazza, A. 1994. Parallels and divergences in the acquisition and dissolution of language. [Series B] *Philosophical Transactions of the Royal Society of London* 346:121–127.

Chesi, C., and A. Moro. 2014. Computational complexity in the brain. In *Measuring Grammatical Complexity*, ed. F. Newmeyer and L. Preston, 264–280. Oxford: Oxford University Press.

Chierchia, G. 2013. *Logic in Grammar*. Oxford: Oxford University Press.

Chierchia, G., and S. McConnell-Ginet. 2001. *Meaning and Grammar*. Cambridge, MA: MIT Press.

Chomsky, N. 1956. Three models for the description of language. *IRE* [now *IEEE*] *Transactions on Information Theory* 2, no. 3 (September): 113–124. Reprinted in R. D. Luce, R. R. Bush, and E. Galanter, eds., *Readings in Mathematical Psychology*, vol. 2, New York: Wiley.

Chomsky, N. 1962. Explanatory models in linguistics. In *Logic, Methodology, and Philosophy of Science*, eds. E. Nagel, P. Suppes, and A. Tarski, 528–550. Stanford, CA: Stanford University Press.

Chomsky, N. 1965. *Aspects of the Theory of Syntax*. Cambridge, MA: MIT Press.

Chomsky, N. 1981. *Lectures on Government and Binding*. Dordrecht: Foris.

Chomsky, N. 2013. Problems of projection. *Lingua* 130 (June): 33–49.

Chomsky, N. 1986. *A Knowledge of Language*. New York: Praeger.

Chomsky, N. 1995. *The Minimalist Program*. Cambridge, MA: MIT Press.

Cinque, G., and L. Rizzi. 2009. The cartography of syntactic structures. In B. Heine and H. Narrog, eds. *Oxford Handbook of Linguistic Analysis*, 51–65. Oxford: Oxford University Press.

Coulomb, Charles Augustin. 1785. Premier mémoire sur l'électricité et le magnétisme. *Histoire de l'Académie Royale des Sciences*, 569–577. Paris: Imprimerie Royale.

Darwin, C. R. 1862. *On the Various Contrivances by Which British and Foreign Orchids Are Fertilised by Insects, and on the Good Effects of Intercrossing.* London: John Murray.

De Bleser, R., and K. Poeck. 1984. Aphasia with exclusively consonant-vowel recurring utterances: Tan-Tan revisited. *Advances in Neurology* 42:51–57.

Dehaene, S. 1999. *The Number Sense. How the Mind Creates Mathematics.* Oxford: Oxford University Press.

Dehaene, S. 2010. *Reading in the Brain: The New Science of How We Read.* London: Penguin Books.

Denes, G. 2011. *Talking Heads. The Neuroscience of Language.* Hove, UK: Psychology Press.

Di Sciullo, A. M. 2005. *Asymmetry in Morphology.* Cambridge, MA: MIT Press.

Domanzki, C. 2013. Mysterious *Monsieur Leborgne*: The mystery of the famous patient in the history of neuropsychology is explained. *Journal of the History of the Neurosciences: Basic and Clinical Perspectives* 22 (1): 47–52.

Du Bois-Reymond, E. 1874. The limits of our knowledge of nature. *Popular Science Monthly* 5:17–32.

Eco, U. 1995. *La ricerca della lingua perfetta: The Search for the Perfect Language.* Trans. B. Laterza. Oxford: Blackwell.

Edelman, G. M. 1988. *Topobiology.* New York: Basic Books.

Edelman, G. M., and G. Tononi. 2000. *A Universe of Consciousness: How Matter Becomes Imagination.* New York: Basic Books.

Embick, D., A. Marantz, M. Yasushi, W. O'Neil, and L. S. Kuniyoshi. 2000. A syntactic specialization for Broca's area. *Proceedings of the National Academy of Sciences of the United States of America* 97 (11): 6150–6154.

Embick, D., and D. Poeppel. 2005. Mapping syntax using imaging: Prospects and problems for the study of neurolinguistic computation. In

Encyclopedia of Language and Linguistics, ed. K. Brown. 2nd ed. Oxford: Elsevier.

Emonds, J. 1976. *A Transformational Approach to English Syntax: Root, Structure-Preserving, and Local Transformation*. New York: Academic Press.

Feynman, R. 1985. *QED: The Strange Theory of Light and Matter*. New York: Penguin Press Science.

Fitch, T., and A. Friederici. 2012. Artificial grammar learning meets formal language theory: An overview. *Philosophical Transactions of the Royal Society* 367:1933–1955.

Frank, R. 2002. *Phrase Structure Composition and Syntactic Dependencies*. Cambridge, MA: MIT Press.

Frank, R., and K. Vijay-Shanker. 2001. Primitive c-command. *Syntax* 4:164–204.

Friederici, A. D., J. Bahlmann, S. Heim, R. I. Schubotz, and A. Anwander. 2006. The brain differentiates human and non-human grammars: Functional localization and structural connectivity. *Proceedings of the National Academy of Sciences of the United States of America* 103:2458–2463.

Friston, K. J. 1997. Imaging cognitive anatomy. *Trends in Cognitive Sciences* 1 (April).

Fux, J. J. 1965. *Study of Counterpoint: From Johann Joseph Fux's Gradus Ad Parnassum*. Ed. and trans. A. Mann. New York: W.W. Norton.

Galilei, Galileo. 1638. *Discorsi e dimostrazioni matematiche intorno a due nuove scienze attinenti la mecanica e i movimenti locali* [Dialogues Concerning Two New Sciences] , vol. 8 (Florence: Barbera, 1890–1909; repr.: 1929–1939, 1964–1968). English translation (1954) by Henry Crew and Alfonso de Salvio, *Dialogues Concerning Two New Sciences*. New York: Dover.

Genty, E., and R. W. Byrne. 2010. Why do gorillas make sequences of gestures? *Animal Cognition* 13:287–301. doi:10.1007/s10071-009-0266-4.

Giraud, A.-L., and D. Poeppel. 2012. Cortical oscillations and speech processing: Emerging computational principles and operations. *Nature Neuroscience* 15:511–517.

Gould, S. J. 1997. The exaptive excellence of spandrels as a term and prototype. *Proceedings of the National Academy of Sciences of the United States of America* 94 (September): 10750–10755.

Gould, S. J. 2002. *The Structure of Evolutionary Theory*. Cambridge, MA: Belknap Press of Harvard University Press.

Gould, S. J., and E. Vrba. 1982. Exaptation—a missing term in the science of form. *Paleobiology* 8 (1): 4–15.

Graffi, G. 1980. Universali di Greenberg e grammatica generative. *Lingua e Stile* [Language and style] 15:371–387.

Graffi, G. 2001. *200 Years of Syntax: A Critical Survey*. Amsterdam: John Benjamins.

Greenberg, J. H., ed. 1963. *Universals of Language*. Cambridge, MA: MIT Press.

Grice, P. 1975. Logic and conversation. In P. Cole and J. L. Morgan, eds., *Syntax and Semantics*, Volume 3: *Speech Acts*, 41–58 . New York: Academic Press.

Hale, K., and J. Keyser. 2002. *Prolegomena to a Theory of Argument Structure*. Cambridge, MA: MIT Press.

Halle, M., and M. Marantz. 1993. Distributed morphology and the pieces of inflection. In *The View from Building 20*, ed. Kenneth Hale and S. Jay Keyser, 111–176. Cambridge, MA: MIT Press.

Hickok, G., and D. Poeppel. 2007. The cortical organization of speech processing. *Nature Reviews: Neuroscience* 8:393–402.

Infeld, L. 1950. *Albert Einstein: His Work and Its Influence on Our World*. New York: Scribner.

Ingram, J. C. L. 2007. *Neurolinguistics. An Introduction to Spoken Language Processing and Its Disorders*. Cambridge: Cambridge University Press.

Jacob, F. [1970] 1974. *La Logique du vivant: Une histoire de l'he're'dité, Gallimard*. [Paris; *The Logic of Life: A History of Heredity*] New York: Pantheon Books.

Jerne, N. 1985. The generative grammar of the immune system. *Science* 229:1057–1059.

Jespersen, O. 1924. *The Philosophy of Grammar*. London: Allen & Unwin.

Jespersen, O. 1928. *An International Language*. London: Allen & Unwin.

Joos, Martin, ed. 1957. *Readings in Linguistics*, 2 vols. Chicago: University of Chicago Press.

Kandel, E., J. Schwartz, T. Jessell, S. Siegelbaum, and A. Hudspeth. 2012. *Principles of Neural Science*. New York: McGraw-Hill Medical.

Katz, J., and D. Pesetsky. January 2011. The identity thesis for language and music. http://ling.auf.net/lingbuzz/000959.

Kayne, R. 1984. *Connectedness and Binary Branching*. Dordrecht: Foris.

Kayne, R. 1994. *The Antisymmetry of Syntax*. Cambridge, MA: MIT Press.

Kayne, R. 2011. Why are there no directionality parameters? In *Proceedings of the 28th West Coast Conference on Formal Linguistics*, 1–23. Somerville, MA: Cascadilla Proceedings Project.

Keller, S. S., T. Crow, A. Foundas, K. Amunts, and N. Roberts. 2009. Broca's area: Nomenclature, anatomy, typology and asymmetry. *Brain and Language* 109 (1): 29–48.

Kelly, B., G. Wigglesworth, R. Nordlinger, and J. Blythe. 2014. The acquisition of polysynthetic languages. *Language and Linguistics Compass* 8 (2): 51–64.

Kennedy, H. 2006. *Peano: Life and Works of Giuseppe Peano*. Concord, CA: Peremptory Publications.

Kubanek, J., P. Brunner, A. Gunduz, D. Poeppel, and G. Schalk. 2013. The tracking of speech envelope in the human cortex. *PLoS One* 8:e53398. doi:10.1371/journal.pone.0053398.

Larson, R. 1988. On the double object construct. *Linguistic Inquiry* 19 (3): 335–391.

Leibniz, G. 1969. *Philosophical Papers and Letters*. 2nd ed. Trans. L. E. Loemker. Dordrecht: Reidel.

Lenneberg, E. 1967. *Biological Foundations of Language*. New York: Wiley.

Lerdahl, F., and R. Jackendoff. 1983. *A Generative Theory of Tonal Music*. Cambridge, MA: MIT Press.

Lesky, A. 1971. *Geschichte der griechischen Literatur*. Munich: De Gruyter Saur.

Levi-Montalcini, R. 1988. *In Praise of Imperfection: My Life and Work*. [Originally published as *Elogio dell'imperfezion*.] New York: Basic Books.

Longobardi, G. 2003. Methods in parametric linguistics and cognitive history. *Linguistic Variation Yearbook* 3:101–138.

Lucretius, Titus Carus. 1977. *On the Nature of Things* [De rerum natura]. Trans. F. O. Copley. New York: Norton. The Latin text is that of Cyril Bailey (Oxford University Press).

Magrassi, L., D. Bongetta, S. Bianchini, M. Berardesca, and C. Arienta. 2010. Central and peripheral components of writing critically depend on a defined area of the dominant superior parietal gyrus. *Brain Research* 1346 (July 30): 145–154.

Magrassi, L., G. Aromataris, A. Cabrini, V. Annovazzi-Lodi, and A. Moro. 2015a. Sound representation in higher language areas during language generation. *Proceedings of the National Academy of Sciences of the United States of America* 112 (6): 1868–1873.

Magrassi, L., G. Cabrini, G. Aromataris, A. Moro, and V. Annovazzi-Lodi. 2015b. Tracking of the speech envelope by neural activity during speech production is not limited to Broca's area in the dominant frontal lobe. Paper presented at the 37th Annual International Conference of the IEEE Engineering in Medicine and Biology Society, Milan.

Manzini, R. 1992. *Locality*. Cambridge, MA: MIT Press.

Marantz, A. 2013. Verbal argument structure: Events and participants. *Lingua* 130:152–168.

Marcus, G., A. Vouloumanos, and I. A. Sag. 2003. Does Broca's area play by the rules? *Nature Neuroscience* 6:651–652.

May, R. 1985. *Logical Form*. Cambridge, MA: MIT Press.

McGuire, P. K., D. A. Silbersweig, R. M. Murray, A. S. David, R. S. J. Frackowiak, and C. D. Frith. 1996. Functional anatomy of inner speech and auditory verbal imagery. *Psychological Medicine* 26 (1): 29–38.

Medawar, P. B., and J. S. Medawar. 1983. *Aristotle to Zoos: A Philosophical Dictionary of Biology*. Cambridge, MA: Harvard University Press.

Mesgarani, N., C. Cheung, K. Johnson, and E.-F. Chang. 2014. Phonetic feature encoding in human superior temporal gyrus. *Science* 343 (6174): 1006–1010.

Miko, I. 2008. Gregor Mendel and the principles of inheritance. *Nature Education* 1 (1): 134.

Monti, M., L. Parsons, and D. Osherson. 2009. The boundaries of language and thought in deductive inference. *Proceedings of the National Academy of Sciences of the United States of America* 106 (30): 12554–12559.

Moro, A. 1988. Per una teoria unificata delle frasi copulari. *Rivista di Grammatica Generativa* 13:81–110.

Moro, A. 1997. *The Raising of Predicates*. Cambridge: Cambridge University Press.

Moro, A. 2000. *Dynamic Antisymmetry*. Cambridge, MA: MIT Press.

Moro, A. 2009. Rethinking symmetry: A note on labelling and the EPP. *Snippets*, July 2009, http://www.ledonline.it/snippets/allegati/snippets19007.pdf. Reprinted in A. Moro (2013). *The Equilibrium of Human Syntax: Symmetries in the Brain*. New York: Routledge.

Moro, A. 2010. *Breve storia del verbo essere. Viaggio al centro della frase*. Milan: Adelphi.

Moro, A. 2011. Clause structure folding and the *wh*–in situ effect. *Linguistic Inquiry* 42 (3): 389–412.

Moro, A. 2012. *Parlo dunque sono. Diciassette istantanee sul linguaggio.* Milan: Adelphi.

Moro, A. 2013. *The Equilibrium of Human Syntax: Symmetries in the Brain.* New York: Routledge.

Moro, A. 2014. How does the dream consciousness/protoconsciousness concept resonate with linguistic ideas and the hypothesis of a Universal Grammar? In *Dream Consciousness,* ed. N. Tranquillo, 167–170. Basel: Springer International Publishing.

Moro, A. 2015. *The Boundaries of Babel: The Brain and the Enigma of Impossible Languages.* 2nd ed. Cambridge, MA: MIT Press.

Moro, A. 2016. *I Think Therefore I Am: Seventeen Snapshots on Language.* New York: Columbia University Press.

Morpurgo Davies, A. 1998: *Nineteenth-Century Linguistics.* Vol. 4, *History of Linguistics,* ed. G. Lepschy. London: Longman.

Musso, M., A. Moro, V. Glauche, M. Rijntjes, J. Reichenbach, C. Büchel, and C. Weiller. 2003. Broca's area and the language instinct. *Nature Neuroscience* 6:774–781.

Newmeyer, F. 2005. *Possible and Probable Languages: A Generative Perspective on Linguistic Typology.* Oxford: Oxford University Press.

Newmeyer, F., and L. Preston. 2014, eds. *Measuring Grammatical Complexity,* 264–280. Oxford: Oxford University Press.

Nourski, K. V., et al. 2009. Temporal envelope of time-compressed speech represented in the human auditory cortex. *Journal of Neuroscience* 29:15564–15574.

Ohno, Susumu. 1972. So much junk DNA. In *Our Genome,* ed. H. H. Smith. 366–370. New York: Gordon and Breach.

Ojemann, G., J. Ojemann, E. Lettich, and M. Berger. 1989. Cortical language localization in left, dominant hemisphere: An electrical stimulation mapping investigation in 117 patients. *Journal of Neurosurgery* 71 (3): 316–326.

Okrent, A. 2009. *In the Land of Invented Languages.* New York: Spiegel and Grau.

Pallier, C., A.-D. Devauchelle, and S. Dehaene. 2011. Cortical representation of the constituent structure of sentences. *Proceedings of the National Academy of Sciences of the United States of America* 108 (6): 2522–2527.

Pasley, B. N., S. V. David, N. Mesgarani, A. Flinker, S. A. Shamma, N. E. Crone, R. T. Knight, and E. F. Chang. 2012. Reconstructing speech from human auditory cortex. *PLoS Biology* 10 (1): e1001251. doi:10.1371/journal.pbio.1001251.

Patel, A. 2003. Language, music, syntax and the brain. *Nature Neuroscience* 6:674–681.

Paul, W., ed. 2013. *Fundamental Immunology.* 7th ed. Philadelphia: Wolters Kluwer.

Penrose, R. 1989. *The Emperor's New Mind: Concerning Computers, Minds, and the Laws of Physics.* Oxford: Oxford University Press.

Piattelli-Palmarini, M. 1989. Evolution, selection and cognition: From "learning" to parameter setting in biology and the study of language. *Cognition* 31:1–44.

Pinker, S. 1994. *The Language Instinct: How the Mind Creates Language.* New York: Morrow.

Poeppel, D. 1996. Neurobiology and linguistics are not yet unifiable. *Behavioral and Brain Sciences* 19 (4): 642–643.

Poeppel, D. 2014. The neuroanatomic and neurophysiological infrastructure for speech and language. *Current Opinion in Neurobiology* 28:142–149.

Pollock, J.-Y. 1989. Verb movement, UG and the structure of IP. *Linguistic Inquiry* 20:365–424.

Price, C. 2012. A review and synthesis of the first 20 years of PET and fMRI studies of heard speech, spoken language and reading. *Neuroimage* 62 (2):816–847.

Pulvermüller, F., and L. Fadiga. 2010. Active perception: Sensorimotor circuits as a cortical basis for language. *Nature Reviews. Neuroscience* 11: 351–360.

Pulvermüller, F., M. Huss, F. Kherif, F. Moscoso del Prado Martin, O. Hauk, and Y. Shtyrov. 2006. Motor cortex maps articulatory features of speech sounds. *Proceedings of the National Academy of Sciences of the United States of America* 103 (20): 7865–7870.

Rabelais, F. 1973. *Œvres complètes*. Vol. 3. Paris: Editions de Seuil.

Raichle, M., and A. Z. Snyder. 2007. A default mode of brain function: A brief history of an evolving idea. *NeuroImage* 37:1083–1090.

Reinhart, T. 1976. *The Syntax of Anaphora*. Doctoral dissertation, MIT.

Rizzi, L. 1990. *Relativized Minimality*. Cambridge, MA: MIT Press.

Rizzi, L. 1997. The fine structure of the left periphery. In *Elements of Grammar*, ed. L. Haegeman. Dordrecht: Kluwer.

Rizzi, L. 2009. The discovery of language invariance and variation, and its relevance for the cognitive sciences. *Behavioral and Brain Sciences* 32:467–468.

Rizzolatti, G., L. Fogassi, and V. Gallese. 2002. Neurophysiological mechanisms underlying the understanding of imitation and action. *Nature* 2:661–670.

Roberts, I. 1988. From rules to constraints. *Lingua e Stile* [Language and style] 23 (3): 445–464.

Roberts, I. 2007. *Diacronic Syntax*. Oxford: Oxford University Press.

Rosen, S. 1992. Temporal information in speech: Acoustic, auditory and linguistic aspects. *Philosophical Transactions of the Royal Society of London* 336:367–373.

Sammler, D., G. Novembre, S. Koelsh, and P. E. Keller. 2013. *Cortex* 49 (5): 1325-1339. doi: 10.1016/j.cortex.2012.06.007.

Saussure, F. de. 1975. *Cours de linguistique générale. Paris: Payot. English translation (1960): Course in General Linguistics*. Rev. ed. London: Peter

Owen. Original French edition published in 1916 and revised editions in 1922 and 1931.

Schroedinger, E. [1944] 1967. *What Is Life and Mind and Matter.* Cambridge: Cambridge University Press.

Shannon, C. E. 1948. A mathematical theory of communication. *Bell System Technical Journal* 27 (July and October): 379–423, 623–656.

Smith, N., and I.-M. Tsimpli. 1995. *The Mind of a Savant.* Oxford: Blackwell.

Swagerman, S. C., E. van Bergen, C. Dolan, E. J. de Geus, M. M. Koenis, H. E. Hulshoff Pol, and D. I. Boomsma. 2015. Genetic transmission of reading ability. *Brain and Language.* S0093-934X(15)00166-2. doi: 10.1016/j.bandl.2015.07.008.

Tattersall, I. 2012. *Masters of the Planet: The Search for Our Human Origins.* New York: Macmillan.

Terrace, H. S., L. A. Petitto, R. J. Sanders, and T. G. Bever. 1979. Can an ape create a sentence? *Science* 206 (4421): 891–902.

Tettamanti, M., H. Alkadhi, A. Moro, D. Perani, S. Kollias, and D. Weniger. 2002. Neural correlates for the acquisition of natural language syntax. *NeuroImage* 17:700–709.

Tettamanti, M., G. Buccino, M. C. Saccuman, V. Gallese, M. Danna, P. Scifo, F. Fazio, G. Rizzolatti, S.-F. Cappa, and D. Perani. 2005. Listening to action-related sentences activates fronto-parietal motor circuits. *Journal of Cognitive Neuroscience* 17 (2): 273–281.

Tettamanti, M., R. Manenti, P. A. Della Rosa, A. Falini, D. Perani, S. Cappa, and A. Moro. 2008. Negation in the brain: Modulating action representations. *NeuroImage* 43 (2): 358–367.

Tettamanti, M., I. Rotondi, D. Perani, G. Scotti, F. Fazio, S. F. Cappa, and A. Moro. 2008. Syntax without language: Neurobiological evidence for cross-domain syntactic computations. *Cortex.* doi:10.1016/j. cortex.2008.11.014.

Thompson, D. [1917] 1961. *On Growth and Form*, ed. J. T. Bonner. Cambridge: Cambridge University Press.

Turing, A. M. 1952. The chemical basis of morphogenesis. *Philosophical Transactions of the Royal Society of London: Series B, Biological Sciences* 237 (641): 37–72.

Vygotsky, L. S. 1986. *Thought and Language*. Trans. A. Kozulin. Cambridge, MA: MIT Press.

Watson, J. D., and F. H. C. Crick. 1953. Molecular structure of nucleic acids: A structure for deoxyribose nucleic acid. *Nature* 171:737–738.

Watson, J. D., T. A. Baker, S. P. Bell, A. Gann, M. S. Levine, and R. Losick. 2014. *Molecular Biology of the Gene*. 7th ed. Boston: Pearson.

Wesson, R. 1991. *Beyond Natural Selection*. Cambridge, MA: MIT Press.

Whitaker, H. A. 1998. History of neurolinguistics. In *Handbook of Neurolinguistics*, ed. B. Stemmer and H. A. Whitaker, 27–53. San Diego: Academic Press.

Wittgenstein, L. 1953. *Philosophical Investigations*. Trans. G. E. M. Anscombe. Oxford: Blackwell.

Yang, C. D. 2003. *Knowledge and Learning in Natural Language*. Oxford: Oxford University Press.

Yang, C. D. 2004. Universal Grammar, statistics or both. *Trends in Cognitive Sciences* 8 (10): 451–456.

Zaccarella, E., and A. D. Friederici. 2015. Reflections of word processing in the insular cortex: A sub-regional parcellation based functional assessment. *Brain and Language* 142:1–7.

INDEX